What People Are Saying About

THE GLORY OF THE TOTAL SOLAR ECLIPSE

"Carefully researched and lovingly created, *The Glory of the Total Solar Eclipse* is a masterclass in writing, with an incredible examination of science, life, and faith that goes beyond the expected and will leave you laughing, crying, and with a newfound joy in your heart. Don't miss this book!"
—Nicole L. Standiford, Computer Scientist, Award-Winning Author

"Amy takes an uncommon approach as she simplifies scientific complexities, connecting the universe to our unique purpose as individuals. Understanding more of the beyond helps to draw our attention upward, and to challenge our alignment with the creator, who surely understands mankind!"
—Bob Unanue, CEO, Goya Foods

"Whether you're a seasoned eclipse chaser or recently experienced your first, *The Glory of the Total Solar Eclipse* is a gift. Amy beautifully weaves the story of Christ's redemption with the phases of the eclipse while interspersing personal stories. She has a gift for making space science accessible to everyone, and each page will remind you of God's love, creativity, power, and hope. You will never look at an eclipse the same way again."
—Jill E. McCormick, writer, speaker, podcaster, Grace in Real Life Podcast

"*The Glory of the Total Solar Eclipse* is a powerful narrative, interweaving real lives of the human experience to an Almighty creator's artsy display. A fantastic display, the eclipse is in constant movement, yet illuminating from every location the need to grab hold of God's path for our lives. As the parents of Britta (Chapter 13), we especially appreciate Amy's heartfelt effort to connect this grand universe to each of our lives, so that we may live as she might say "in totality" of God's design. You'll be blessed by enjoying this book!
—**Elsa and Charles Ezell,** Meant to Soar Foundation

"A wonderful and exceptionally well-written and researched story of the perfect total solar eclipse, interwoven with the beauty of intelligent design, and God's promise of salvation through Jesus Christ."
—**Rob Thorpe,** Software Program Manager

"Step into the radiant pages of *The Glory of the Total Solar Eclipse,* where every subtle detail of this celestial dance unfolds. Amy lovingly invites readers to open their hearts and spiritual eyes to unlock ethereal truths found in this natural wonder. For those of us looking toward the heavens for inspiration, this book is a must read.
—**Mario and Danielle Lopez,** Author and Co-Author of *How I Know*

"Amy Burgin has crafted an overview of the Total Solar Eclipse that introduces us to the grace, wonder, and faithfulness of God. Her words, from a solidly biblical perspective, compare and contrast the glory of a solar eclipse with the story of the world and my own place in it—a tiny speck in the universe and yet a valued, loved, and cherished child of God. Thanks to Amy's clear writing, scientific principles stick and spiritual concepts unfold. You will want a copy of *The Glory of the Total Solar Eclipse* to go with each pair of viewing glasses you share!
—**Mandy Pallock,** Author of *The Question Habit*

THE GLORY OF THE TOTAL SOLAR ECLIPSE

Discovering God's Extravagant Love in Its Path

Amy Marie Burgin

Shribble Publishing
SAN ANTONIO, TEXAS

Shribble Publishing
The Glory of the Total Solar Eclipse: Discovering God's Extravagant Love in Its Path.

Available online at amyburgin.com

The Glory of the Total Solar Eclipse / Amy Marie Burgin
ISBN: 979-8-9884789-3-5
Library of Congress Control Number: 2024903588
Science & Math / Science & Religion / Inspirational
First Edition

All emphasis in Scripture citations has been added by the author.

Edited by Jill E. McCormick
Front cover design: Joy Burgin Fish
Back cover design: Mandy Pallock
Interior design: Mandy Pallock
Illustrations: Joy Burgin Fish
Illustrator photo: Christina Ramirez
Author photo: Joy Burgin Fish

This book is typeset in Adobe Caslon Pro and Lao MN.

Printed in the United States of America.

Published in association with Shribble Publishing.
shribblepublishing.com

To those who hunger and thirst
for beauty and goodness on Earth.

For further information, visit amyburgin.com.

CONTENTS

AUTHOR'S NOTE

Multiple kinds of eclipses occur, such as partial solar, annular solar, lunar, and more. This book focuses on Total Solar Eclipses, so when you read the word "eclipse" without any other description, I am referring to the Total Solar Eclipse.

PREFACE

I witnessed the Total Solar Eclipse that crossed America in 2017. This extraordinary event exceeded my every expectation, overwhelming me with a deep sense of awe, a response shared by most people.

This book explores the wonderment of the eclipse. I believe the creator of the universe wants to speak to us through all of creation, including the Total Solar Eclipse. Experiencing the eclipse fills us with a profound mix of fear, wonderment, surprise, magnificence, mystery, and more—all at once. It's almost as though we can feel the eclipse speaking to the deepest part of ourselves; we feel its voice vibrate the very marrow of our bones. I believe if we sit still and listen, we can know what it's telling us. We can put its speech into words.

What if the eclipse is telling us that evil can never eclipse good? That we have a creator whose plan for us is perfect and good? What if Totality's crown speaks of the death of a king who lives forever? What if the Diamond Ring Effect speaks of a promise kept? What if the eclipse tells the story of our world?

The eclipse testifies to something glorious. Our bodies testify in agreement. In this book, we explore science, stories, and faith in our quest to uncover the mechanics, message, and mystery of the Total Solar Eclipse and what it means for us.

CHAPTER ONE

MY TRIP TO TOTALITY

From Texas to Kentucky

"Hello?" I answered my phone in the parking lot of our local grocery store.

My mother-in-law greeted me in a breathless hurry. "Hi, Amy! Karl and Debbie are coming in for the Total Solar Eclipse. It only happens every 100 years. The best place to see it is Kentucky. I'm booking rooms on the base in Fort Campbell right now. Do you guys want to join us?"

Even though I had never heard about the eclipse and knew nothing about it, I trusted her ability to point me to the wonderful things in life. Furthermore, the excitement in her voice was contagious. "I'd absolutely love to. What are the exact dates?"

"Well, the eclipse is on August 28, but it's a pretty long drive over there, so we're going to leave about a week in advance, and then I'm going to book the rooms for August 25 to 29."

"Great! Let me check my calendar!" With my mother-in-law on speaker, I opened my calendar app. "Oh," I responded flatly, deflated and full of disappointment. "August 28 is the first day of school."

Is It Worth the Trip?

My mind raced between extremes. My oldest daughter was about to start her senior year of high school and couldn't care less about the first day of school. I knew beyond a shadow of a doubt she would want to see the eclipse. She takes after her dad. Spontaneity is one of their strengths.

However, my youngest daughter, who would start her sophomore year of high school, takes more after me. Anxiety about performing well would prevent her from skipping the first day of school with all its essential first-day instructions and introductions. No school year could be successful if she missed those.

But I pictured the people and realized all of Jon's living brothers would be together. With one in Switzerland and one in Kansas, having them all in one place would make for another rare event!

My mind raced back to my youngest daughter, though, and I pictured her face and her fear, and I landed on a decision.

"We can't go," I said. I chose the required routine of life, convincing myself we couldn't go. What kind of mother would I be if my children missed the first week of school?

I drove home with a van full of groceries and put all the cold stuff in the fridge before researching the Total Solar Eclipse for the first time in my life. I read the Moon will cover the Sun for a minute or so, and it will be like night in the middle of the day. I also discovered that Fort Campbell, Kentucky—almost 1,000 miles away—was indeed an optimal viewing location. Wavering on my decision again, I wondered: *What kind of mother would I be if my children missed the Total Solar Eclipse and a chance to take a 1,000-mile family road trip?*

No, the thought of missing the first week of school churned up feelings of devastation. Although I felt saddened at the thought of missing the eclipse, I didn't feel devastated. After all, I had only heard about it that day. And we can look at the night at night. We didn't need to see night in the middle of the day. Still, something deep in my bones told me it would be quite a thing to see night come out at noon.

One Month Later

One month later, I was on the phone again with my mother-in-law. She chatted about the dates my brother-in-law Karl and his wife Debbie would be in town. I realized she was talking about the week before school started, not the first day of classes.

"Wait!" I stopped her. "What day is the eclipse?"

"Oh, I think it's August 24 or 21."

Sitting at home in front of my PC, I googled the answer while on the call with her. "It's August 21! Mom! Is it too late to get us a room?"

"I don't know, honey. Everything is booking up fast. You guys can go?"

"Yes, can you call right now to check for us?"

"Sure, I'll call you back."

Overjoyed when she called back to tell me she had booked a room for our family of five, we began to prepare for our trip. The girls were excited about "seeing night at noon," as I explained it.

The drive from San Antonio, Texas, to Fort Campbell, Kentucky, is about 14 hours. I researched the journey and created a little itinerary to make stops at all the best places along our path. We packed our five-day eclipse adventure with drive time and touristy visits, stopping at the Crater of Diamonds State Park, Dunbar Cave, and a replica of the Parthenon until we met up with family on Sunday night, the day before the eclipse.

On the day of the eclipse, we settled under a pavilion Mom Burgin had reserved at Fort Campbell several hours before Totality. Except for a few people walking by here and there, our family group was alone in a wide open space at this military base. Jon's brother CJ pulled Mom and Dad Burgin's RV from Texas to Kentucky, which they'd packed with food for a grand picnic. We brought a game of Bananagrams and played a few rounds while we waited for the Moon to get close to the Sun. CJ passed out the protective glasses he'd ordered for all of us. Wishing to be mostly alone and experience the phenomenon without the noise of human chatter, I set up my immediate family's folding chairs on the side of the pavilion opposite Jon's parents and brothers.

So It Began

The partial phase began at 11:56:44 a.m., at which time we all stepped out from under the pavilion to view it with our protective glasses. It proved uninteresting at the start. The Moon took a tiny bite out of the Sun's image, but you could hardly notice. We resumed chatting under the pavilion and stepped out every five or 10 minutes to put on our paper glasses and look up again. The Moon inched in front of the Sun, but we noticed little change on Earth. Our interest grew when the Sun's shape slowly changed from a circle to a thinning sickle. As Totality (when the Moon completely covers the Sun) neared, crescents of light surprised us by dancing on the ground amid the shadows of leaves.

It proved uninteresting at the start.

Minutes before the moment of Totality, we settled into our folding chairs under the sky and looked up constantly. Until this point, the progression of the eclipse perfectly matched a simulation we had watched on YouTube. When only a thin, ever-narrowing orange sliver of the Sun shone in the sky, I expected the Moon to cause the Sun to disappear totally, as the simulation suggested.[1] Instead, a brilliant, dazzling diamond ring in the sky shocked my senses. We all gasped. Staring in wonderment, we removed our paper glasses as the diamond ring transformed into a crown—the only moment it's safe to look at the Sun with your naked eye. My eyes fixated on a living crown in the sky, a ring of

1 Fischer, Eric. "Total Solar Eclipse U.S.A. August 21st, 2017 (Simulation)" *Youtube,* uploaded by Eric Fischer, 22 April 2015, www.youtube.com/watch?v=vzJqeyxye_E.

white fire shifting, undulating, and reaching out from behind the Moon in all directions. I had never known the Sun this way before.

My husband tried to take a picture with his phone, but the camera wouldn't focus. He began to grumble. Unable to bear a single negative sound spoiling this sacred, awe-filled moment, I whispered, my eyes glued to the living crown, "Don't try. Millions of people around the world are capturing photos. Just enjoy it."

A brilliant, dazzling diamond ring in the sky shocked my senses.

Our 12-year-old son, Freedom, sat between us, his eyes also sealed to the living crown. He echoed my words with a slow, quiet, amazed voice, "Yeah, millions of people are taking photos right now."

The sky did not darken as much as I expected, but a few stars twinkled during this time. Crickets chirped, and I quietly repeated the words, "Oh my God." Unable to take my eyes off the light show and unaware of the 360° sunset effect, I missed the change in the horizon. I later learned that Totality creates the beauty of sunrise and sunset simultaneously by blanketing the horizon in all directions with twilight's violet and pink colors.

For over two minutes and 29 seconds, we all sat there, eyes fixed on this living crown, only to be dazzled again by a second bright, white diamond ring, our cue to veil our eyes once more with a protective film. The Moon continued its appointed journey across the Sun as the white diamond ring quickly turned to a dark orange, hair-thin sickle. We lingered in our chairs with our faces up as the sickle thickened and freed us from its trance. My husband, children, and I exchanged smiles and looks of wonder with widened eyes as we marveled at the astonishing beauty.

What in the world just happened? I wondered. Nothing had prepared me for such glory.

Eclipse chaser Dr. Kate Russo described what I felt in her

book *Being in the Shadow: Stories of the First-Time Total Eclipse Experience:* "I knew technically what was about to happen—the Moon would block all light from the Sun, casting us into a dark shadow. But I was completely unprepared for the immersive experience . . . It was the most awe-inspiring nature event I had ever seen."[2] Like Dr. Russo, I was utterly unprepared. I've seen meteor showers, the cotton crops of West Texas as far as my eyes could see, the redwoods of California, the Black Forest of Germany, wild baboons and hippos in Africa, the waterfalls and geysers of Yellowstone National Park, the sunset on the ocean in the Gulf of Mexico—all of these brought me joy and wonder, but none left me as awestruck as the Total Solar Eclipse.

So It Ended

It would take the Moon another hour and a half to move out from between us and the Sun. Like the wait for Totality, we found this phase uninteresting. The mountaintop experience was over, and it was time to climb down. Even though people had snacked all morning, the schedule declared it was now time to eat lunch. My husband, children, and I made our way back to the pavilion where the rest of the family had already gathered. Everyone quickly agreed the eclipse was amazing, and all faces were redirected to the task of setting out our barbeque lunch. I was good at following their lead and did so while quietly pondering the eclipse in my heart. I didn't know what to say—more than anything, I wished to sit alone with nature a little longer, but that would have appeared rude.

Debbie, Jon's sister-in-law, said she thought she captured a good photograph. She did. Debbie sent me a copy of her photo, an excellent image of Totality. I pinned it to the bulletin board at my desk.

2 Russo, Kate. *Being in the Shadow: Stories of the First-Time Total Eclipse Experience.* Belfast, Northern Ireland, UK, Melinda Martin, 2017. p. 1.

After the meal, we made small talk, took a group photo, carried all the leftovers back to the RV, and started washing dishes. Just like that, the eclipse passed over, and we were back to the mundane: to dish towels and discussions about weather and traffic.

Even though our group quickly returned to the ordinary chores of life, I couldn't put the Total Solar Eclipse out of my mind. No. Totality remained with me, demanding my attention daily. Questions bombarded my thoughts as I put the groceries away, shuffled the laundry, or watched spaghetti noodles boil on the stove. Why did that experience bring about such emotion? I felt like the Creator was speaking to me, but what was he saying?

So...What's the Point?

The point is that the eclipse draws us in, and we're curious to know more. Perhaps you're still processing your first Total Solar Eclipse experience. Or maybe you're a seasoned eclipse chaser addicted to the experience. Or someone who is preparing for their first time in Totality. Regardless, you're drawn to and compelled by the eclipse—inclined at least to pick up this book.

You're not alone. The total eclipse of the sun compelled me to invest time, energy, and resources exploring various aspects of the eclipse and eventually writing this book. I even pursued a new career supporting space science due to my rare rendezvous with the Sun and Moon. And we're hardly alone—the eclipse has strongly affected people across the globe from ancient times to today, as we'll discover in Chapter Two.

Most people exclaim, "Wow! Oh my God!" when they see a Total Solar Eclipse, and that's what I hope you'll find in this book: stories and ideas that make you respond with those same words. I'll compare the Total Solar Eclipse with God's extravagant love and beauty through Jesus Christ, as evidenced in the story of our world.

In Chapter Three, I'll share my response to Totality, including the launch of my new career supporting space science and the moment of epiphany when I discovered how each phase of the eclipse maps to the story of the world.

In Chapter Four, I hope we'll become friends as I'll share a glimpse of my childhood to my young adult years. We'll ponder our littleness amid the backdrop of our vast universe and examine our need for salvation. We'll give ourselves permission to wonder if there may be help outside ourselves.

In Chapter Five, we'll consider our special place in the universe, often called the Goldilocks Zone, and how humanity's understanding of our special place has evolved throughout history. We'll wonder together what it means to find ourselves in a rare and favorable place of such a vast space.

In Chapter Six, we'll examine the precise mechanics required for a Total Solar Eclipse. We'll consider how good designs don't come about by chance but are created on purpose by diligent and skilled designers. We'll allow ourselves to be in awe of the Great Creator whose designs not only thrill us but also serve as the foundation for our own creations. We'll pick up the courage to dream of a Great Creator who designed the eclipse on purpose, and we'll awaken our curiosity to wonder what that purpose might be.

In Chapter Seven, we'll examine the beginning of our world in a light you may not have seen before. We'll discover how God created us and our world with great pleasure, uncovering his desire to be with us and share all his beauty without holding anything back.

Chapters Eight through Twelve compare each phase of the eclipse to the story of the world. First Contact, when the Moon first begins to obscure the Sun, reminds me of the Fall, when Adam and Eve displayed their distrust in God, elevating their ideas above his and separating from him.

Second Contact, the first Diamond Ring Effect, points to the

First Covenant God made with humanity: Follow my laws, and it will go well with you. Although no person could measure up to his rightness, the First Covenant brought forth an acute awareness of our sin and our need for a savior. The first diamond ring also points to God's promise to send a savior to fix the problem.

Totality, the climax phase of the eclipse, maps to the climax of our world, the promise fulfilled through Jesus Christ, King of kings on Earth, crucified, buried, and raised to life.

Third Contact, the second Diamond Ring Effect, expresses the New Covenant God offers to humanity: Trust in me, and I will give you a pure heart, reconciling you to myself. I will restore you and everything back to its perfect beginning.

Fourth Contact, when the Moon continues its journey until it no longer takes up space between us and the Sun, reminds me of God's growing kingdom on Earth—the gospel's spread unto the end of the age.

In the final chapter, we'll study the path of Totality, the only place on Earth where a person can experience the beauty of a Total Solar Eclipse. We'll compare this to the narrow path of Jesus Christ, the only way to come to God. The effort required to enter the path of Totality is worth the trip. Stepping into the narrow path of Jesus is even more worth it.

In every chapter, we'll connect stories, science, and faith in ways that allow us to explore the eclipse and discover the beauty of God. Together, we'll find the Creator is telling us that evil can never eclipse good, and our failures can never eclipse God's love for us.

Evil can never eclipse good, and our failures can never eclipse God's love for us.

CHAPTER TWO

AWESTRUCK THROUGH THE AGES

From Ancient China to Modern America

After the Total Solar Eclipse, one of my first questions was, *Is anybody else as awestruck as I am?* It didn't take much research to discover I wasn't alone. People have been astonished by solar eclipses across the globe and through the ages. From ancient China to modern America, the eclipse's wordless speech has strongly compelled people's emotions, often leading them to take action.

The Devouring Dragon in Ancient China

The ancient Chinese thought a dragon in the sky tried to swallow the Sun, so people shouted and banged pots and drums to scare the serpent off.[1] It may sound strange to think the Moon a dragon, but Annie Dillard makes a keen observation in her classic essay, "Total Eclipse." She writes, "You may read that the Moon has something to do with eclipses. I have never seen the Moon yet.

1 Petruzzello, Melissa. "The Sun Was Eaten: 6 Ways Cultures Have Explained Eclipses". *Encyclopedia Britannica,* 1 Aug. 2017, http://www.britannica.com/list/the-Sun-was-eaten-6-ways-cultures-have-explained-eclipses. Accessed 7 May 2023.

You do not see the Moon."[2] Annie is right. You don't see the Moon in a Total Solar Eclipse because the brilliance of the Sun hides it. Yes, the Moon is there, moving closer and closer to the Sun until it completely covers it, but your eyes never see it coming. It appears as if some invisible thing simply eats away at the Sun.

You don't see the Moon in a Total Solar Eclipse because the brilliance of the Sun hides it.

Although the commoner, banging his pots and pans, may not have understood the cause of an eclipse, ancient astronomers did and could precisely predict the majestic event as they did in Ancient Assyria.

The Death of a King in Ancient Assyria

Ancient astronomers in the cradle of civilization perceived the coming of an eclipse as a warning of a calamitous event, specifically the death of a king.

Astronomical Diaries, a collection of ancient fragments of astronomical observations as old as circa 1800 B.C. up to 20 B.C., describe upcoming eclipses as omens that foretold a king's death and the extreme actions people took to protect him from his pending doom. A substitute king was chosen to take the real king's place until the danger passed. The surrogate could have been a criminal or an ordinary civilian. Dressed in royal clothes and placed on the royal throne, he pretended to rule until the time to carry the punishment meant for the king came, and he was put to death as the scapegoat. With the bad news of the eclipse brought to fruition, the true king safely resumed his place on the throne.[3]

2 Dillard, Annie. "Total Eclipse." The Best American Essays of the Century, edited by Joyce Carole Oats, co-edited by Robert Atwan, Harper Collins Publishers, 2000, pp. 477-489.

3 Parpola, Simo, et al. *Letters from Assyrian Scholars to the Kings Esarhaddon and Assurbanipal,* Eisenbrauns, 2007.

While Assyrians viewed the eclipse as a sign of death, other ancients saw the eclipse as a message to make peace.

A Call for Covenant in Ancient Lydia and Media

The ancient historian Herodotus authored *The History of Herodotus* in 440 B.C. Thanks to Project Gutenberg, a free online library, we can read a translation. In it, Herodotus wrote about a multi-year war between the Lydians and Medes under the kingship of Alyattes and Cyaxares, respectively. He describes a particular battle in the sixth year of the war: "Just as the battle was growing warm, the day was on a sudden changed into night. This event had been foretold by Thales…The Medes and Lydians, when they observed the change, ceased fighting and were alike anxious to have terms of peace agreed on." Herodotus goes on to describe their taking of oaths backed by the arranged marriage of their children and the "slight flesh wound in their arms, from which each sucks a portion of the other's blood."[4] NASA maps this event to the Total Solar Eclipse of May 28, 584 B.C.[5]

It's not only the ancients who thought eclipses came with a message; musings from the Middle Ages show that people of that time considered an eclipse as a display of God's sorrow.

The Death of a King in the Middle Ages

William of Malmesbury, a prolific writer, monk, and well-respected historian of the Middle Ages, authored *Deeds of the Kings of the English.* William implied the Total Solar Eclipse of

4 Herodotus. *The History of Herodotus.* Translated by G. C. Macaulay. E-book, Project Gutenberg, 1 December 2008. www.gutenberg.org/files/2707/2707-h/2707-h.htm. Accessed 21 May 2023.

5 NASA. "Solar Eclipses of Historical Interest" *NASA Goddard Space Flight Center Eclipse Web Site,* 28 September 2009, eclipse.gsfc.nasa.gov/SEhistory/SEhistory.html. Accessed 21 May 2023

A.D. 1133 foretold the fate of King Henry's death and expressed nature's grief. Malmesbury wrote:

> The providence of God, at that time, bore reference in a wonderful manner to human affairs: for instance, that he [King Henry I] should embark, never to return on alive... the elements manifested their sorrow at this great man's last departure. For the sun on that day, at the sixth hour, shrouded his glorious face, as the poets say, in hideous darkness, agitating the hearts of men by an eclipse.[6]

The idea of humankind finding messages in Total Solar Eclipses did not stop with the Middle Ages but continued into early modern times.

Contemplation in Early Modern Times

A reporter at *Virginia Argus,* a semiweekly newspaper published in Richmond, Virginia, in the early 1800s, wrote this about people's response to the Total Solar Eclipse that occurred on Monday, June 17, 1806:

> The great eclipse...forcibly brought to our recollection those lines of Cato—'There is a God above us, All nature cries aloud through all her works.' Even the mind of mere *curiosity* and *listlessness* felt a *sensible* impression from the awfully magnificent scene...But to awakened and reflective minds, it was truly a 'feast of reason and a flow of soul'—it irresistibly led them up to the highest order of contemplation—to admire the 'works of an Almighty hand'—and feel the very *littleness* of human ability.[7]

6 Giles, J. A. *William of Malmesbury's Chronicles of the Kings of England.* 28 December 2015. https://www.gutenberg.org/cache/epub/50778/pg50778.txt . Accessed 05 October 2023.

7 Virginia Argus. [N. 1275.] (Richmond, Va.), 09 July 1806. Chronicling America: Historic American Newspapers. Lib. of Congress. https://chroniclingamerica.loc.gov/lccn/sn84024710/1806-07-09/ed-1/seq-3/

The article describes a deep sense of awe toward the creator and a heightened sense of smallness in humanity. But the annular eclipse that crossed America approximately two decades later compelled an enslaved man to lead a murderous rebellion.

A Violent Rebellion in Early Modern Times

On February 12, 1831, an annular solar eclipse passed over the southern states of America. Most articles refer to this as a solar eclipse, and it's easy for readers to assume it was a *Total* Solar Eclipse. Media advertised the eclipse in advance, preparing observers to be awestruck, but when the event occurred, observers were sorely disappointed.[8] Annular (meaning ring-shaped) eclipses are not as mesmerizing as Total Solar Eclipses because you only see a ring of fire. You don't see a stunning diamond ring or a mesmerizing corona because the Moon doesn't fully cover the Sun.

Nat Turner, a 31-year-old man enslaved in Virginia, saw the annular eclipse as a sign to commence a murderous rebellion. According to his confession, recorded by attorney Thomas R. Gray, Nat states, "And on the appearance of the sign (the eclipse of the sun last February) I should arise and prepare myself, and slay my enemies with their own weapons."[9] Nat continues to describe the execution of his plan, now known as Nat Turner's Rebellion, which began with the late-night murder of the Travis family of five, from father to infant.

Nat assumed others must have also gotten the same message from the eclipse. When asked about an insurrection in another state, Nat declared he knew nothing about it. But when Gray gave the appearance of disbelief, Nat responded, "I see sir, you doubt my word; but can you not think the same ideas, and strange

8 Masur, Louis P. "Eclipse." 1831, *Year of Eclipse,* Hill and Wang, New York, 2002, pp. 3–8.
9 Turner, Nat. *The Confessions of Nat Turner.* E-book, Project Gutenberg, 12 March 2005. www.gutenberg.org/files/15333/15333-h/15333-h.htm. Accessed 21 May 2023.

appearances about this time in the heaven's *[sic]* might prompt others, as well as myself, to this undertaking."[10]

Eclipses continue to speak to the heart of humanity in the 21st century.

Eclipse Chasing in Modern Times

Even those of us in the modern day are awestruck by eclipses. Travel across the globe is possible like never before, enabling people to pursue and collect Total Solar Eclipse experiences.

Astronomer Jay Schneider, who has seen at least 35 Total Solar Eclipses, planned to take a photo of his first eclipse but, like me, found himself paralyzed by the diamond ring and crown. "I was thunderstruck," he said. "I couldn't move. I didn't take a single photo. I was mesmerized. It froze me right to the spot, and changed my perception of what I was going to do with the rest of my life."[11]

Even those of us in the modern day are awestruck by eclipses.

Eclipse observer Troy Anders, interviewed in the video, *Eclipse 2017: One Nation Under The Sun | NPR,* traveled 13 hours to see the eclipse in Kentucky. It compelled him to permanently mark his body with his first tattoo: a picture of the eclipse above the place and date of his experience, "Hopkinsville, KY 8.21.2017."[12]

A Total Solar Eclipse will not transform every observer into an eclipse chaser or compel every observer to get a tattoo, but listen to any video of eclipse footage, and you almost always hear people in the background repeatedly declare, "Oh my God! Oh my God!"

10 Ibid.

11 *Totality: The American Eclipse,* Written, directed, and produced by Eryl Cochran, Bird Rock Productions, 2017.

12 "Eclipse 2017: One Nation under the Sun | NPR." YouTube, *YouTube,* 27 Aug. 2017, www.youtube.com/watch?v=VAEUYM4Een4. Accessed 05 October 2023.

Something about the eclipse commands our attention and speaks to our bones.

So...What's the Point?

The point is the heavens, including the Total Solar Eclipse, declare a message. Will we make space and time to listen? As I write this, I sit outside a coffee shop in Texas, and the traffic roars by. A white work pickup truck stops at the red light with music blaring. The city roars around me, breaking the silence of knowing. Little opportunity exists to stop and listen to what the Earth and stars say to us. When a Total Solar Eclipse crosses our path, however, it demands those in its shadow to hush, and it ushers in our attention. We look up in awe and wonder: *What do the heavens declare?*

The point is the heavens declare a message. Will we make space and time to listen?

From ancient times to today, people have been awestruck and often compelled by the messages they find in a Total Solar Eclipse. People have believed it was evil's attempt to swallow the good, foretold the death of a king, or is a message of covenant. In modern times, eclipses created a whole new people group: eclipse chasers. Eclipses have uniquely compelled societies around the globe and in different periods.

But what did the eclipse mean for me? How was God speaking in the eclipse of 2017? How does an eclipse tell the story of the world?

I had to find out.

CHAPTER THREE

MY RESPONSE TO TOTALITY

Asking God

After witnessing the 2017 Total Solar Eclipse in Fort Campbell with my family, I couldn't stop thinking about it. Why did it stir up such emotion? Was it because it was so rare? So other-worldly? Why did the eclipse come with so much power to leave me awestruck? Was it simply the sheer beauty of it?

One morning, as I sat at my kitchen table, I stopped wondering and asked God instead. I asked, "Why, God? Why does the eclipse move me so? Why does it leave people through the ages and across the globe awestruck? What is significant about it?"

Immediately, I heard a surprising answer in the stillness of my mind. The Lord shared, *Because it speaks of my covenant and Kingship. And because it tells the story of the world. It tells your stories.*

The simple, surprising answer filled my mind with whirling thoughts of the story of the world as seen through the Christian lens, as well as my testimony and the testimonies of my Christian friends.

In the Moon's attempt to eclipse the Sun, I saw humanity's

choice to distrust God. I saw my own choice to separate from God and go my own way when deceived by evil. I saw evil's attempt to eclipse that which God blessed and made good. I saw evil's attempt to rule over the whole world by slowly taking over our hearts, making us slaves to ungodly beliefs, jealousy, anger, bitterness, isolation, pride, sexual immorality, and greed.

In the first diamond ring, I saw God's response to evil's intent in the Old Testament. Rings represent promise and covenant. I saw God's promise that he would ensure Eve would bear the living victor, who would crush the head of this eclipsing evil (Genesis 3:15). I saw the first covenant God made with humanity through the commandments given to the nation of Israel.

In the living crown that left me stunned, I saw the King of kings, Jesus Christ, on Earth, crucified and raised to life, the living victor promised in the first diamond ring.

I saw God's new covenant through Jesus in the second diamond ring. I saw his promise that whoever believes in Jesus can be reconciled to God.

As the Moon continued its journey across the Sun, ultimately ending the eclipse, I saw God's work to make everything new: Heaven, Earth, and you and me. I saw God's work of sanctification in the hearts of his believers, freeing them from the slavery of evil desires, ungodly beliefs, hurts, and wounds. I saw God's faithfulness to his promise that no evil can eclipse his marvelous plan for his people.

> In the Total Solar Eclipse, I saw that no evil can eclipse God's love for us but only reveals his love more deeply.

In the narrow path of Totality, the only way to observe a Total Solar Eclipse, I saw God's call to all humanity: *Come to me through my Son—he's the only way. I love you. I want to give you the fullness of joy for which I created you. I want to abide with you forever.*

In the Total Solar Eclipse, I saw that no evil can eclipse God's love for us but only reveals his love more deeply.

That simple response in the stillness of my mind and the whirling thoughts that followed left me awestruck. It filled me with a similar awe and wonder I experienced during the 2017 eclipse.

If this light show in the sky symbolizes the story of our creator's extravagant love manifested in the measures he has taken to ensure no evil eclipses his marvelous plan for us, then no wonder it speaks to our very bones.

At that moment, my interest in Total Solar Eclipses grew exponentially, and I began learning as much as I could about them.

I turned to research and contemplation. The eclipse became more meaningful with each day of consideration. I learned the images I saw in the sky, the rings and crown, were scientific terms used to describe facets of the eclipse. The first and second diamond rings map to eclipse phases, and scientists refer to the phenomenon as the *Diamond Ring Effect.* The moment of Totality reveals the Sun's *corona,* which means *crown.* Across the globe, rings represent covenants, and crowns represent authority and reign. I continued to study the eclipse and how it maps to the story of the world. My research even led me to a new career in space science.

A New Career in Space Science

I have a master's degree in mathematics with an emphasis in computer science. (I hope that sentence didn't cause your eyes to glaze over). Immediately upon graduation, I became a software developer. After my third child was born, I felt God led me to quit my career (which I thoroughly enjoyed) and become a full-time homemaker. I obeyed, assuming I would return to work after my youngest started kindergarten; however, I never felt the freedom to return, so I continued to dedicate myself to family and volunteer work for the next 13-and-a-half years.

In 2018, we entered what I call "the expensive season," when our oldest started college with her two younger siblings following directly behind her. Wishing to launch our children well after graduating high school, with the gift of used cars, college degrees, and no debt, I eased into work as a part-time math professor at our local college. The pay was insufficient to cover all the bills ahead of us. I knew I needed full-time work but wasn't sure who I wanted to work for. At this time, while researching the eclipse, I checked out a documentary from the library, *NOVA: Eclipse over America.*

In this documentary, I learned that Southwest Research Institute (SwRI), a top-tier research organization based in my hometown of San Antonio, had a branch in Boulder, Colorado. At this location, they installed cameras on the front of NASA jets and flew with the Moon during the 2017 eclipse to take pictures of the Sun's corona.[1] Right then and there, sitting on the brown leather loveseat in front of the television, I heard God's quiet voice: *Apply for a software development job at SwRI in space exploration.*

Almost 20 years prior, immediately after college graduation, I worked for SwRI, but not in space science. Knowing the Boulder branch put the cameras on the NASA jets, I imagined no work related to space exploration was performed in my hometown. Nevertheless, I obeyed the still, small voice. I paused the documentary, walked upstairs to my computer, found the SwRI website, and searched their job listings. They had approximately two dozen software developer positions open in San Antonio, and one of those involved working in space science. I applied for that one.

I wondered if I could step back into the field after a 13-year pause in my career. Within a couple of weeks, I received a phone

1 Gorst, Martin, and Martin Gorst. *NOVA: Eclipse Over America.* Windfall Films Ltd., 2017.

call from Rob, who shared that he would like to bring me in for an interview; however, the next day, I received a phone call from Brad.

"I understand you spoke with Rob yesterday about coming in for an interview. I'm not so sure that's a good idea, so Rob suggested I give you a call. I see you have a long gap in your career."

I appreciated Brad's bluntness because I, too, was concerned about being out of the field for so long. We spoke frankly about the gap.

Then he asked, "Why, out of the dozens of open software development positions here, did you only apply for this one?"

"It's the only one I want," I said simply without explaining my divine directive.

Brad agreed to bring me in for a lengthy interview, and I got the job. The 2017 eclipse led me to a new career in writing software in support of space missions. Within two years, I found myself as the software lead for the Juno Science Operations Center, a suite of software products managing data from the spacecraft Juno. This spacecraft embarked on a journey to Jupiter in 2011, entered the gas giant's orbit in 2016, and regularly sends data and images of Jupiter to Earth.

Stepping into a new career in space exploration after being a homemaker for over a decade may sound impressive to some, but it's nothing compared to knowing Jesus.

How Can We Know Jesus?

Jesus poses an important question, perhaps the most important question on the planet. He asks, "Who do you say I am?" (Luke 9:18). In his book, *Mere Christianity,* C.S. Lewis ponders possible answers to this question and concludes Jesus must be either a liar, lunatic, or Lord. Lewis explains that we can't accept Jesus as a great moral teacher but not accept his claim to be God. "He would either be a lunatic—on the level with the man who says he is a poached egg—or else he would be the Devil of Hell."[2] But how can we know?

To know Jesus, we must put a measure of trust in him and experience him for ourselves. It's impossible to know the wonder of a Total Solar Eclipse without experiencing it. I can explain it all day, but my words will fall short, and you will not understand it for yourself. It's impossible to know another person without spending time with her. And it's impossible to know who Jesus is if we're not in a relationship with him. We can't experience the awe of a Total Solar Eclipse without entering its path, and we can't know Jesus without putting a measure of trust in him.

It's impossible to know the wonder of a Total Solar Eclipse without experiencing it.

So...What's the Point?

The point is Jesus is more beautiful than the Total Solar Eclipse. You can know him, and it's important to me that you don't miss him.

My journey to the path of the Total Solar Eclipse led me on a journey to space and beyond in ways I never imagined. I've been changed by the eclipse, inside and out. And while I don't mean for you to pursue a career serving space exploration as I did, I long to invite you on the journey with me.

Those who have experienced an eclipse from the path of Totality want to share its extraordinary beauty and encourage all

2 Lewis, C. S. "Book 2. What Christians Believe Chapter 3. The Shocking Alternative." *C. S. Lewis Classic Collection: Mere Christianity*, Harper One, New York, NY, 2001, pp. 47–52.

to enter. Likewise, those who know Jesus well want everyone else to know him, too. My primary response to the Total Solar Eclipse is to tell you that Jesus is much more awesome than the eclipse. Please come with me, enter his narrow path, abide with me there, look up, and be awestruck. I can't bear for you to miss such a beautiful Person, such a beautiful God.

I want to tell everyone who doesn't personally know Jesus but may have experienced the Total Solar Eclipse, "Hey! Did you love the Total Solar Eclipse? Guess what? This eclipse experience can't compare to walking with Jesus! Come enter his narrow path with me and allow me to point you to his beauty!"

And I want to tell everyone who has put a measure of trust in him but cannot find his beauty under death's shadow, "Friend, Jesus is concerned with your hurt and grief. Come with me to find his face in this place."

And I want to tell everyone who does know Jesus, enjoys his beauty in the day-to-day, and seeks to soak in as much of him as she can, "Hey! Isn't he amazing?! Let's gaze upon his beauty together—look how beautiful he is through this lens of the Total Solar Eclipse!"

You picked up this book, so I believe you have been called out of darkness to know Jesus for the first time or to know Jesus more than you already do. Will you walk with me and learn with me in the coming chapters? We'll connect stories, science, and faith in ways that ultimately point us to the beauty of God.

CHAPTER FOUR

A SPECK IN NEED OF SALVATION

Hints of Help from Outside Ourselves

Shortly after starting work in space science, my friend Luke introduced me to the Pale Blue Dot. On Valentine's Day, 1990, the Voyager 1 spacecraft took a photo of Earth from the other side of Pluto. This photo has become known as the Pale Blue Dot because, in it, that's precisely how Earth looks. She looks like a dot of light blue, a single pixel, in a band of red sunlight amid a black backdrop.

In his book of the same name, Carl Sagan described Earth as "a lonely speck in the great enveloping cosmic dark." He goes on to declare, "In our obscurity, in all this vastness, there is no hint that help will come from elsewhere to save us from ourselves." Sagan presumes there is no creator; if there is, he does not care to help us.

When the photo reached Earth, I was an invisible speck on this pale blue dot, a junior in high school living alone in a small West Texas town in the house of my long-deceased great-grandmother. I didn't feel like I had a trajectory and mission. Instead, I felt like a random, purposeless asteroid hurtling through space after my parents inherited and moved into that house when I turned 12.

Childhood

Before we moved to West Texas, our family moved at least every summer to follow my dad's work as a crane operator. By sixth grade, I had attended eight different schools. It didn't bother me even though I experienced bullying in some schools and never made a friend in others. I enjoyed learning, observing people, and coming home to the refuge of my parents and poodle at the end of each day. The three of us and our little dog made a good team. My dad was our Sun, my mom was his orbiting Earth, and I was her orbiting Moon.

Every morning, my parents woke before our side of Earth faced the Sun. They didn't turn on the radio or television, but my dad belted out the chorus of two songs repeatedly: "Honky-Tonk Man" by Johnny Horton and "Folsom Prison Blues" by Johnny Cash.

My mom brewed coffee and packed his lunch. His songs often woke me, and I would stumble out of bed to watch him lace up his work boots while he sang. He was happy, handsome, 6'2", and wore starched Wrangler jeans and a brown, long-sleeved, snapped Western shirt. My mom was beautiful, quiet in the morning, one foot shorter than my dad, and in pajamas. They kissed goodbye. He left like a sunset before sunrise. She stayed like the Earth beneath my feet, and I sleepily went back to bed until it was time for school.

When Dad came home each evening, Mom served him a glass of iced tea, and I unlaced his work boots while she served a meal complete with meat, starch, and vegetables. After dinner, my parents watched television and I retreated to paper, pens, pencils, and crayons in my room. Each spin of Earth was the same, and we were all that mattered.

When Great Grandma Goss died, my dad followed the work, but my mom and I stayed in the inherited house because my parents thought I needed the stability of a single school.

Adolescence

From ages 13 to 16, my dad was absent from our daily lives. My parents were still married, and Dad worked to provide everything we needed, but we lost our unity when we no longer lived together. With my dad out of sight and working over eight hours away in South Texas, my gaze changed, and my life orbited three main interests: the flute, a boyfriend, and looking like I walked out of an MTV video every day. My mom supported my hobbies.

I dominated the band's first chair flute position with arrogance, which meant I was the best flutist at our school, one of the best in the region, and highly competitive. My mom chaperoned every band trip and provided piano accompaniment to those who participated in solo competitions.

When I fell in love at 13, I became obsessed with my boyfriend. At this tender young age, he and I spent too much time alone doing things we shouldn't have, and I lost my innocence. I had learned everything I needed to know from Mom's collection of paperback romance novels.

My mom went to cosmetology school, which I thought was amazing because she could give me the most beautiful blonde highlights in school. I wore ripped jeans, dark blue eyeliner, and white or pink lipstick, which was fine—as long as my dad didn't know. Dad wouldn't have approved of white lipstick.

Shortly before I turned 16, the boyfriend, who I was utterly soul-tied to, broke my heart. When I gained the independence that comes with a driver's license, Mom and Dad briefly separated. My mom lived elsewhere, and that's when I started living alone in

Great Grandma's house. I never attempted suicide, but day after day, the thought bombarded my mind: *If I kill myself and leave the right note behind, my parents will get back together.*

I tied myself to another boy, and we had ample space to be alone for entire nights. We did everything together. Sometimes, we drank together. Since we were underage, he picked up homemade alcohol from a bootlegger in Lubbock. Whenever we drank, I drank as much as I could as fast as I could until I couldn't do anything at all. One night, I overdosed on the bootlegger's concoction and couldn't lift myself from the floor.

I remember one thing from that night—looking down on myself from the ceiling with my boyfriend kneeling beside me. The next day, my boyfriend admitted he had considered taking me to the hospital the night before, concerned I might die from alcohol poisoning. He opted against it to avoid potential trouble for our drinking. I stopped drinking after that.

My mom and dad reconciled, and we all moved to South Texas in the middle of my senior year. A week passed before school started, and I reconnected with my love of reading. We owned an old encyclopedia set, and I read the book of C from cover to cover. Camels had three stomachs, and cicadas lived underground for as long as I had been alive—17 years! Students warmly welcomed me the first day I went to school, but I experienced my first panic attack when I came home. It made no sense. It was like some crawling black beast came out of nowhere and overtook my body. My mother held me that night, absorbing the terrorizing energy of unexplained fear.

I stubbornly refused to return to the new school the following day, and no one made me go. I moved back to West Texas to live by myself again in Great Grandma's house and graduate from my previous high school. This time, I drove to Nanny's almost every weekend. My dad's mom, Nanny, lived an hour away. She cooked for me, respected me, loved me, listened to me, and sang about

Jesus the rest of the time. Her love for Jesus started turning my gaze toward God. Since I had no idea what to do next, I attended college in my hometown.

College

My boyfriend, the same one I drank with, broke up with me and went to work with his father out of state. That's when I met Jon, my professor at Texas Tech University. I sat in the front row of his class. He didn't seem that much older than I was, maybe a handful or two. His adventurous eyes and mischievous grin gave him a youthfulness. At 5'6", he was a little short for my type, but he was taller than me, which was enough. He had a cute twinkle in his eye and reminded me of Michael J. Fox, who almost every girl from the 1980s had a crush on at some point. He moved with energy when he taught and had big muscles that flexed nicely when he pointed at his terrible handwriting on the board.

Completely enamored with him, I had dreamed visions of who he was. For example, he didn't dress very well, so I imagined he came from an impoverished family—probably the first in his family to go to college like me. Then again, because he smiled a lot and had bulging biceps, I figured he came from a farming family and earned them naturally by stacking hay bales on a trailer. I imagined he loved Jesus because surely, I couldn't have a crush on anyone who didn't.

I did not imagine that my professor also had a crush on me. When I turned in the final exam, he asked if I could come to his office later that day at four o'clock. I agreed, assuming he wanted to share a scholarship opportunity. When I arrived in his office, he stood up from his desk and walked around to stand before me,

"You made an A on the final exam and in the class."

"That's great," I smiled.

"I've turned in all the grades," he continued.

I looked at him.

"All the grades have been turned in, and the class is officially over."

I blinked.

"The class is officially over," he repeated.

I blinked more.

"Would you like to go out on a date with me?"

My stomach flipped. "Yes!"

We dated for six months. I learned that Jon was only eight years older than me, was not the first to attend college, came from a wealthy family, was far from being a farm boy, and got angry every time I talked about Jesus.

Nearing college graduation, I guessed marriage was the next step in life. I didn't know how to choose a spouse, so I prayed a simple prayer one day. I said, "Lord, if I'm supposed to marry Jon, let him call me next. If I'm supposed to marry my high school sweetheart, let him call me next. If I'm not supposed to marry either, please have my dad call me next." I sat in my bedroom at Great Grandma Goss's house and stared at the white phone with its long, squiggly cord until it rang.

"Hello?"

"Hi!" Jon answered.

A Wedding (Or a Funeral?)

My parents came into town the day before my wedding in May 1995. The morning of the big day, my dad stayed in bed longer than usual. With my wedding dress on a hanger, hooked on my first two fingers, and draped behind my right shoulder, I walked by the open door of his bedroom. He had been crying. "Just load up all the gifts in the truck, and let's get the hell out of here!" he called out. He was half joking and heartbroken, not ready for his little girl to be grown up. I don't remember responding as I left.

I didn't have the fortitude to deal with his feelings on top of mine. The day before my wedding, I struggled so much with worry that I went to a doctor, hopeful he could help me stop my anxiety-induced need to run to the toilet and empty my bowels multiple times an hour. As I write this, almost 28 years later, the image of a prescription bottle in my chest of keepsakes flashes through my mind. I hurry to the closet, open the trunk, find the translucent, dark orange bottle buried beneath old letters and cards, and read the label.

> BLACKSTOCK, AMY M**
>
> TAKE 1 OR 2 TABLETS FOUR TIMES A DAY FOR NERVOUS STOMACH
>
> 05/12/95

My memories have not forsaken me. My grandmother described my wedding as "not like Amy's wedding, but more like Amy's funeral." I called Nanny yesterday to ask her if she remembered saying that. She doesn't, but she figured she said it because she wasn't ready to give away her little blonde-headed girl. I figure Nanny said this because none of the tears shed that day were joyful except Jon's.

When my dad walked me down the aisle, I cried. He gave me away and left me on a stage next to this man I barely knew in front of my Uncle Doug, who officiated. I don't remember what Uncle Doug said, but I remember how I felt. I felt terrified. A wedding picture testifies, depicting my red, wet, anguished face with eyes looking not at Jon nor my uncle but at the sky. I pleaded silently, "Oh God, we need your help to make this work. What am I doing? Who exactly is this man? Please be with us. You just gotta make this work." I needed help. I needed salvation.

Salvation from What?

We've come a long way in astrophysics since Carl Sagan's passing in 1996. Given what we've found so far through space exploration, I think Sagan might stand by his quote, "[T]here is no hint that help will come to save us from ourselves."

"From ourselves"—these may be the key words. Living under the Cold War's fearsome threat of nuclear destruction, Sagan concerned himself with the villainous rulers of Earth. He wrote, "Think of the rivers of blood spilled by all those generals and emperors so that, in glory and triumph, they could become the momentary masters of a fraction of a dot." I think of men like Stalin, Mussolini, and Hitler, who believed themselves larger than they were—men who shed innocent blood to rule over a plot of land so small that it's indistinguishable on a pale blue dot.

Sagan recognized our need for salvation from ourselves.

Sagan recognized our need for salvation from ourselves, and I also recognize that need. Not only do I see our need for salvation from criminal behavior on a global scale, such as that exhibited by terrorism, genocide, and war, but I also recognize it on a personal scale. I need salvation from enemies that attack me personally, some enemies that seem to live inside me. From adolescence to marriage, I needed salvation from many enemies:

the enemy of division that broke up my family.

the enemy of suicide that said others would be better off if I ceased to be alive.

the enemy of worthlessness that lied to my high-school self and assured me I was only a random asteroid with no real purpose, plan, or trajectory.

the enemy of insecurity that led me to seek attention through heavy makeup and heavy drinking.

the enemy of fleshly desire that stole my innocence and broke my heart.

the enemy of anxiety that at any moment terrorized my body and stole my courage.

These are only a few. Many enemies bullied me. Enemies continue to attack me today, but I have found a help and shield that changes everything.

Help Is There

Over 13 years would pass before Jon and I could discuss Jesus without conflict. Twenty-five years after we tied the knot, we watched our digitized wedding ceremony on the flat-screen TV mounted on our living room wall. We understood few words because of poor audio quality, but we heard my Uncle Doug paraphrase Psalm 121:1, "I lift my eyes up to the hills—where does my help come from? My help comes from the Lord, the maker of heaven and Earth." A sense of awe settled inside me as I thought of how the Lord had faithfully answered my request at the wedding altar with my red, anguished face and wet eyes looking up at the sky.

God had helped us.

So...What's the Point?

The point is we do need help outside ourselves, and help is there. Though both Sagan and I set our gaze on the heavens, Sagan focused, looked, and listened with a different set of eyes and ears than I do. Where he searched with various spacecraft instruments, I searched with prayer. Perhaps that's why Sagan saw no hints of help, and I see many. I see a hint of help when I experience prayers answered beyond a terrified bride's imagination.

I see a hint of help when darkness tries to shroud the Sun in the Total Solar Eclipse but fails miserably. Light dazzles us instead with a diamond ring and a crown—but we cannot know unless we look up in the path of Totality. So it is with prayer. Darkness descends on our lives but fails in the end. God dazzles us—we can know it if we abide in his narrow path and look up.

> Where he searched with various spacecraft instruments, I searched with prayer.

We're all in need of help. We need rescue from more than the criminal rulers of Earth, like Hitler, Stalin, and Mussolini. We need rescue from enemies like pain, grief, exhaustion, division, fear, anxiety, comparison, isolation, depression, selfishness, and more. Will you lift your face, red, wet, and anguished as it may be, with mine in prayer? I have found help is there.

CHAPTER FIVE

OUR PLACE IN THE UNIVERSE

Goldilocks and the Three Sweet Spots

All six of us (my husband, three adult children, and my daughter's boyfriend) piled into the van several hours before sunrise to ensure we wouldn't miss any part of the sea turtle hatchling release at Padre Island National Seashore. Darkness grew around us once we reached Park Road 22 of the island, where city lights ceased. Upon arrival, we made our way across the lit boardwalk and then stood on the dark beach where rangers and volunteers prepped the release sites and kept the growing crowd informed, corralled, and out of the way.

Venus sparkled and rose from the east over ocean waters. Orion stood sideways over the waves, and Jupiter kept watch directly overhead. Within an hour, a wide, muted, pink ribbon appeared over the east horizon, announcing the coming entrance of the Sun. I gathered my four young adults for a photo in front of the sunrise. With one in a skirt, one with a top-knot, and all arms holding hands stretched up toward heaven, their gray silhouettes looked like paper dolls.

In the distance, rangers raked three large areas of sand and surrounded each one with cones and caution tape. A plump, enthusiastic ranger with a wide gait and a red headlamp ran toward the crowd and invited us to cross the invisible boundary that held us back. Some began running, each seeking a front-row seat to one of the three release sites.

Rangers circulated the crowds, teaching us about the Kemp's ridley sea turtles. The scientific community did not know the location of their nesting site until 1947, when a pilot took amateur video footage showing approximately 40,000 females nesting on a beach in Mexico. Poachers stole ninety percent of the eggs. In an effort to preserve them, conservationists moved many eggs to South Padre Island in the 1970s to establish a protected nesting site.

Today, people comb the Texas seashore during nesting season to find their buried eggs. When found, half are moved to protected areas of the beach where cars are not allowed. The other half are incubated in a controlled environment and then released on shore. Critically endangered, the Kemp's ridley sea turtle population now sits between 7,000 to 9,000. Some say baby sea turtles use the Moon as a guide to find the ocean, but no ranger mentioned the Moon.

Rangers brought out a Styrofoam™ cooler of baby sea turtles while four volunteers formed a square around them with a string of colorful flags fixed to four tall PVC pipes. Two other volunteers, each carrying a tall flag, stood between the constructed square and ocean waves. When seagulls flew near, the volunteers raised and waved their flags to shoo them away.

Rangers removed the baby sea turtles from the cooler and set them in the square facing the sea. Some turtles quickly took off toward the water, some didn't do much, and a few turned around and slowly started toward the dunes. The Rangers turned those back around to ensure they went in the correct direction. If a turtle stood still too long, rangers would help it along by moving it a few feet closer to the water's edge. My son and my daughter's boyfriend made bets on which baby sea turtle would enter the ocean first (my son won). Within 45 minutes, they all made it to the water, where waves swept them in, forcing them to swim for the first time. Their little heads bobbed up and down.

If conservationists create special, protected places to ensure the survival of sea turtles, what might it mean if we find ourselves in a special, protected place in the universe?

Are We Special?

Baby sea turtles are small enough to fit in your hand. We're small in the universe, specks on a pale blue dot in need of salvation. We long for salvation from pride, greed, bitterness, loneliness, anger, anxiety, and many more enemies. We see hints of help outside ourselves by recognizing answered prayers. We can also find a hint of help by recognizing our special place in the universe.

Although the universe is vast beyond our measure, and although it does not revolve around us like most before Galileo Galilei believed, we have found ourselves residing in a unique, protected place inside it. It's reasonable for us to dream, therefore, that if we reside in a special place, perhaps we are positioned here by a special God who loves us very much.

It's reasonable for us to dream, therefore, that if we reside in a special place, perhaps we are positioned here by a special God who loves us very much.

Learning Our Place in the Universe

Consider the history of humankind's understanding of our place in the universe. Initially, people believed the Sun orbited our planet. It's not surprising. When we stand outside and watch the Sun rise in the east and set in the west, it certainly appears that way. From my perspective, I'm standing still, and the Sun orbits me. But the Sun doesn't revolve around me or planet Earth as astronomers and scientists first believed.

Ptolemy's Diagram

Ptolemy, an Egyptian astronomer and mathematician, was the first person to publish a diagram of the universe around A.D. 150. He showed Earth in the center of the universe with the Sun and the planets of our solar system orbiting us. Around our solar system, he drew stars encased in a rotating canopy.[1]

The Sun doesn't revolve around me or planet Earth as astronomers and scientists first believed.

What does such a picture suggest about us? What does it say about us if we are the center of the universe? What does it mean if a fixed shield of stars binds us, and what do we think resides outside that starry shield? Are we bothered by the limitation of a starry fence? Must we break through it to find what lies beyond?

This belief that the Sun and all other heavenly bodies orbit Earth is known as the geocentric view. How special we must seem in this view—the entire universe revolves around us!

We have a running joke in our house. I stumble into a conversation and assume everyone is talking about me. My husband or kid will look at me and grin, "Not everything is about you, Mom."

1 Wile, Jay L. "MODULE #1: A Brief History of Science." *Exploring Creation With General Science*, 2nd ed., Apologia Educational Ministries, Kendallville, IN, 2008, p. 10.

I laugh and say, "Well, it should be!" I love believing life revolves around me, just as people first loved to believe all heavenly bodies revolved around the Earth. People loved the idea so much that discoverers of the opposing truth were afraid to speak for fear of backlash.

Copernicus and a Revolving Earth

Approximately 500 years ago, a Polish priest, mathematician, and astronomer named Nicolaus Copernicus embraced an unpopular idea. He theorized that Earth rotated daily on its own axis, causing the Sun to appear to rise in the east and set in the west. Do you remember learning about Earth's rotation? It wasn't an easy concept for me to grasp as a child.

I didn't believe it because I couldn't feel myself spinning. I knew what it was to spin until you were dizzy: your legs gave way, and your body spilled onto the floor, feeling sick and wonderful all at the same time. Ancient Greek philosopher Aristotle, who lived in the 4th century B.C., also didn't believe it. He experienced no constant wind blowing over the Earth's surface. When he threw a ball straight up in the air, it didn't land behind him.[2] The idea of Earth rotating seems to contradict what we see with our eyes.

Copernicus embraced another unpopular idea. Copernicus theorized that Earth revolved around the Sun, a heliocentric view. For fear of criticism, he waited almost three decades before he published these ideas in his book *On the Revolutions of the Heavenly Spheres.*[3] Perhaps Copernicus was wise to wait. He couldn't prove his theories, and in 1600, sixty-five years after he published, Italian scientist Giordano Bruno was burned at the stake for embracing the heliocentric view and propagating the

2 Riebeek, Holli. "Planetary Motion: The History of an Idea That Launched the Scientific Revolution." *NASA Earth Observatory,* NASA, 7 July 2009, earthobservatory.nasa.gov/features/OrbitsHistory. Accessed 22 Nov 2023.

3 Copernicus, Nicholas, *Six Books on The Revolutions of the Heavenly Spheres,* translated by Charles Glen Wallis, 1543 archive.org/details/OnTheRevolutionsOfTheHeavenlySpheres/mode/2up. Accessed 27 May 2023.

ideas of an infinite universe and multiple worlds.[4,5] However, Galileo's telescope proved the geocentric view false, thus paving the way for the acceptance of a Sun-centered system.

Galileo's Telescope

In 1610, Galileo created what may have been the most powerful telescope of his time.[6] Using it to examine the night sky, he recorded what he saw. What did he see? Galileo saw what looked like four stars moving around Jupiter. Although this did not prove Earth orbits the Sun, it did provide a counterexample to the geocentric view. At least four heavenly bodies did not orbit Earth.

Humanity's Earth depends on and revolves around an immovable fountain of light.

Galileo published his findings in a little book, *The Sidereal Messenger*, also known as *The Starry Messenger*.

Sidereal (pronounced sai-DEE-ree-uhl) is not a word we come across often. Starry, celestial, glittering, luminous, and spangled are some synonyms. *Sidereal* comes with the special connotation of thinking about stars that are distant, like the ones you can't see with your naked eye.[7] For the first time, our eyes could reach the distant stars. The distant stars carried a clear message: *Not everything revolves around humanity on Earth.*

Galileo was put on trial for heresy and finished his life under

4 Riebeek, Holli. "Planetary Motion: The History of an Idea That Launched the Scientific Revolution." *NASA Earth Observatory*, NASA, 7 July 2009, earthobservatory.nasa.gov/features/OrbitsHistory. Accessed 22 Nov 2023.

5 Britannica, The Editors of Encyclopaedia. "Giordano Bruno summary". Encyclopedia Britannica, 14 Oct. 2003, www.britannica.com/summary/Giordano-Bruno. Accessed 22 November 2023.

6 Van Helden, Albert. "Galileo." *Encyclopaedio Brittanica*, 17 May 2023, www.britannica.com/biography/Galileo-Galilei/Telescopic-discoveries. Accessed 27 May 2023.

7 Galilei, Galileo, Johannes Kepler. *The Sidereal Messenger of Galileo Galilei*. E-book, Project Gutenberg, 19 June 2014, www.gutenberg.org/cache/epub/46036/pg46036-images.html. Accessed 27 May 2023.

house arrest because of his discoveries, but his opponents could not stop the truth. Humanity's Earth depends on and revolves around an immovable fountain of light. What does such an idea suggest about humanity? How does it impact our significance?

The Heliocentric View and Our Significance

As telescopes grew more powerful and the knowledge of humanity expanded, it became unswervingly evident that not everything revolves around us.

We learned Earth is one of eight planets that orbit the Sun. The Sun and all the heavenly bodies that orbit it comprise our solar system. **Earth is not the center of our solar system.**

Our solar system resides in a galaxy, a gargantuan collection of approximately 100 million stars we call the Milky Way. **Our solar system is not in the center of our galaxy.**

Our galaxy dwells inside the universe, also referred to as the cosmos, the collection of everything that exists: galaxies, matter, energy, space, time, and whatever else we find. **Our galaxy is not in the center of the universe.**

For some time, it seemed we held no special place in the cosmos. This knowledge led many to conclude we are nothing special outside our own ideas about ourselves. Carl Sagan promoted this idea in his book, *Pale Blue Dot.* Sagan starts the chapter titled "A Universe Not Made For Us" with a summary of the heliocentric view. He ends the chapter with these words:

> The significance of our lives and our fragile planet is then determined only by our own wisdom and courage. We are the custodians of life's meaning. We long for a Parent to care for us, to forgive us our errors, to save us from our childish mistakes. But knowledge is preferable to

> ignorance. Better by far to embrace the hard truth than a reassuring fable. If we crave some cosmic purpose, then let us find ourselves a worthy goal.[8]

But a heliocentric view does not cause my mind to leap to the same conclusion. When I commute from work to home down Bandera Road, I drive under sixteen traffic lights that do not revolve around me; they do not turn green when I approach. Has my significance diminished? Of course not. As the traffic lights and our obedience to them serve to get commuters home safely, the Sun and our orbit around it serve to provide the necessary conditions for life. Our significance does not depend on the universe revolving around us.

Our significance does not depend on the universe revolving around us.

We look and discover a universe large and vast, one whose edges we have yet to find, as Giordano Bruno speculated over 400 years ago. We find stars galore and think, *My goodness, life like ours, rich and complex, must flourish everywhere! With all the possibilities out there, surely other planets like ours exist with all the elements required to host life as we know it.* But today's astrophysicists agree that the position in which we are placed is so unique that the chances of it occurring randomly are preposterously slim.[9] As we continue to make more discoveries about our place in the universe, we have come full circle. Though we are not the center of our solar system, galaxy, or universe, we have come to learn we do reside in a very special place after all.

8 Sagan, Carl. "A Universe Not Made for Us." *Pale Blue Dot,* The Random House Publishing Group, 1994, pp. 54–55.

9 Smethurst, Becky. "Aliens Probably Exist." *Space at the Speed of Light: The History of 14 Billion Years for People Short on Time,* Ten Speed Press, 2020, p. 96.

The Goldilocks Zone

Our place is so extraordinary and so right for life as we know it that scientists use the term the Goldilocks Zone to describe it. You likely know the story of Goldilocks, but maybe you don't, so I'll give you a quick summary of this English fairy tale.

Out on a walk in the woods one day, a young girl named Goldilocks stumbled into the house of a family of bears. Now, the bears were out on a walk of their own, so she found the house empty. In the bear house, she found a table with three bowls of porridge, which was incredibly convenient because she was hungry after her walk. She took a bite of Papa Bear's porridge, but it was too hot. She took a bite of Mama Bear's, but it was too cold. She took a bite of Baby Bear's, and it was just right, so she gobbled it all up!

She discovered their lounge chairs. Papa Bear's was too hard, Mama's too soft, but Baby Bear's chair was just right. Snooping around, she discovered their beds! After trying all three, she found Baby Bear's was perfect for a nap.

Baby Bear's porridge, chair, and bed were made just right for someone just like Goldilocks. Similarly, our planet, solar system, and galaxy are made just right for life like ours:

1. Our Earth and Moon are in the just-right spot in the solar system.
2. Our solar system is in the just-right spot in the galaxy.
3. Our galaxy is in the just-right spot in the universe.

Our Earth and Moon's Just-Right Spot in the Solar System

First, we abide in an extremely unique solar system, one with a Sun that is not too small, not too large, and not too volatile in its energy production. Although there are billions and billions of stars out there, not many are right for hosting life like ours.

British Astrophysicist Becky Smethurst says there is only one out of a trillion stars out there similar to our Sun.[10] If our Sun were cooler, Earth would need to be much closer to achieve the correct temperatures, but to get closer means boiling off our atmosphere and becoming tidally locked.

We abide in an extremely unique solar system, one with a Sun that is not too small, not too large, and not too volatile in its energy production.

If Earth were tidally locked, one side of Earth would face the Sun for extended periods and scorch. The other side would freeze in the darkness.[11] A day would be equal to a year—the amount of time it would take Earth to orbit the Sun would be the same amount of time it takes for Earth to do a full rotation on its axis. If you're anything like me, this can be difficult to visualize.

A little science experiment devised by my brilliant husband can demonstrate. But maybe you don't feel like getting up to perform the experiment because you're curled up in a cozy spot reading this book. If that's the case, skip the experiment and look at the illustration in Figure 1. There, you will find me tidally locked about my husband. Enamored, I revolve around him, and although I spin myself, my face is fixed like flint on him.

If Earth were tidally locked around the Sun, Earth would be inhospitable to life because only one side of Earth would ever be bathed in sunlight. Our Earth may not be the center of our solar system, but it sits inside it in the perfect spot for life, not too close and not too far from our just-right Sun. The position and size of our Moon also play an important role in supporting life.

10 Ibid., p. 92.
11 Ibid., pp. 89-92

SCIENCE EXPERIMENT #1

Understand What It Means to Be Tidally Locked

Figure 1

Materials

- A friend or object to play the part of the Sun.
- Yourself to play a tidally locked planet about the Sun.

Procedure

Face "the Sun" and slowly walk around "the Sun" keeping your face toward "the Sun" for one revolution around "the Sun."

Conclusion

Did you notice you also rotated your own body one full rotation to keep your face toward "the Sun?" You were tidally locked about your friend, "the Sun."

Our Moon is tidally locked around Earth and is unusually large compared to the moons of other planets in our solar system. The Moon is 27% the size of Earth. In contrast, Triton, the solar system's second-largest moon relative to its parent planet, comprises only 8.8% of Neptune's size.[12,13] Scientists believe the Moon's gravity may spin the liquid iron inside Earth's core, thus serving a critical role in the generation of Earth's magnetic field.[14] The magnetic field shields Earth from solar winds (dangerous particles emanating from the Sun). This protective armor prevents our planet from becoming a barren wasteland like Mars, thought to have had its atmosphere and water stripped away by solar winds.[15] Additionally, the Moon's size and position stabilize Earth's tilt, which is vital for balanced weather patterns.

If Earth's tilt were significantly larger, the Northern Hemisphere would experience searing temperatures half the year when it faced the Sun and viciously cold temperatures the other half of the year when it faced away from the Sun. If Earth's tilt were significantly smaller, Earth's wind patterns would prevent the wide distribution of rain needed to support life across the globe.[16]

This perfect positioning and sizing of Earth, Moon, and Sun lend themselves to a habitable zone within our solar system. Further, our solar system's position in the Milky Way resides in a protected space hospitable for life.

---

12 Dobrijevic, Daisy, and Tim Sharp. "How Big Is The Moon?" *Space.com,* Space, 28 Jan. 2022, www.space.com/18135-how-big-is-the-moon.html. Accessed 29 Dec. 2023.

13 "Solar System Sizes - NASA Science." *NASA,* NASA, Oct. 2023, science.nasa.gov/resource/solar-system-sizes/. Accessed 29 Dec. 2023.

14 Todd, Iain. "Is the Moon Maintaining Earth's Magnetism?" *BBC Sky at Night Magazine,* 31 Mar. 2018, www.skyatnightmagazine.com/news/is-the-moon-maintaining-earths-magnetism. Accessed 29 Dec. 2023.

15 Gough, Evan. "We Might Know Why Mars Lost Its Magnetic Field." *Universe Today,* 11 Feb. 2022, www.universetoday.com/154461/we-might-know-why-mars-lost-its-magnetic-field/. Accessed 30 Dec. 2023.

16 Gonzalez, Guillermo, and Jay W. Richards. *The Privileged Planet: How Our Place in the Cosmos Is Designed for Discovery.* Regnery Publishing Inc., 2004. p. 4-7.

Our Solar System's Just-Right Spot in the Galaxy

Not every place in the galaxy is safe for life. For example, we would not want Earth placed near a supermassive black hole named Sagittarius A* (pronounced "Sagittarius A star"), which resides in the heart of the Milky Way. It's an object so dense (over 4 million times the mass of the Sun)[17] that its gravitational force draws in and destroys everything that gets close enough to it, including stars and light.[18] It's easy to see we want to stay away from its event horizon, where we would get sucked in. But the event horizon is not the only dangerous place.

Not every place in the galaxy is safe for life.

Surrounding Sagittarius A* is the greatest concentration of stars in the galaxy.[19] This section of the galaxy can be seen from Earth in the constellation Sagittarius. Inside the constellation, a group of stars form the shape of a teapot. At the spout of the teapot sits Sagittarius A*. The concentrations of stars around it appear like steam coming from the spout.[20] It's important that we're not in the teapot's steam because stars emit dangerous radiation.

Stars are nuclear fusion power plants flying through space. Simply put, the pressure inside their cores is so enormous that multiple hydrogen nuclei are fused together to create a helium nucleus, resulting in the emission of extreme amounts of energy,

17 "Supermassive Black Hole Sagittarius A*." *NASA,* NASA, 29 Aug. 2013, www.nasa.gov/image-article/supermassive-black-hole-sagittarius/. Accessed 23 Nov. 2023.

18 "Black Holes - NASA Science." *NASA Science,* NASA, 22 May 2023, science.nasa.gov/astrophysics/focus-areas/black-holes/. Accessed 22 Nov. 2023.

19 Ward, Peter D., and Donald Brownlee. "Habitable Zones in the Galaxy." *Rare Earth: Why Complex Life Is Uncommon in the Universe,* Copernicus Books, New York, NY, 2004, p. 27.

20 McClure, Bruce, et al. "Teapot of Sagittarius Points to Milky Way Center." *EarthSky,* 12 Aug. 2023, earthsky.org/favorite-star-patterns/teapot-of-sagittarius-points-to-galactic-center/. Accessed 23 Nov. 2023.

light, and dangerous radiation like X-rays and gamma rays.[21] To be situated amid a plethora of stars is to be destroyed by high amounts of radiation. We don't have to worry about that because our solar system is positioned far away from the dense cluster of stars found at the center of the galaxy.[22] Furthermore, our solar system mostly avoids the galaxy's spiral arms.[23]

Not all galaxies are shaped the same. Ours looks like a flat round dish if viewed from the side and appears to look like it has spiral arms when viewed from above. The activity and radiation in the areas of the spiral arms densely crowded with stars and asteroids can be inhospitable to life. Our solar system resides in a local void on the edge of a small, partial spiral arm called the Orion Arm.[24] We travel around our galactic center at a rate similar (but not equal) to that of the spiral arms, thus avoiding the danger of crossing them very often (only once about every 200 million years).[25,26] We're situated in the perfect spot of our galaxy (at least for the next 100 million years), and we're in a nice spot of the universe as well.

We're situated in the perfect spot of our galaxy.

21 Wile, Jay L. "An Introduction to Astrophysics." *Exploring Creation with Physical Science,* 2nd ed., Apologia Educational Ministries, Kendallville, IN, 2013, pp. 397–401.

22 Ward, Peter D., and Donald Brownlee. "Habitable Zones in the Galaxy." *Rare Earth: Why Complex Life Is Uncommon in the Universe,* Copernicus Books, New York, NY, 2004, p. 27.

23 Abe, Shige. "Galactic Habitable Zones." *Astrobiology at NASA: Life in the Universe,* NASA, astrobiology.nasa.gov/news/galactic-habitable-zones/. Accessed 23 Nov. 2023.

24 "The Milky Way Galaxy - NASA Science." *NASA,* NASA, Oct. 2023, science.nasa.gov/resource/the-milky-way-galaxy/. Accessed 28 Dec. 2023.

25 Abe, Shige. "Galactic Habitable Zones." *Astrobiology at NASA: Life in the Universe,* NASA, astrobiology.nasa.gov/news/galactic-habitable-zones/. Accessed 23 Nov. 2023.

26 Randall, Ian. "Earth's Crust Grew Faster When Our Planet Passed through the Milky Way's Spiral Arms, Study Suggests." *Physics World,* 8 Sept. 2022, physicsworld.com/a/earths-crust-grew-faster-when-our-planet-passed-through-the-milky-ways-spiral-arms-study-suggests/. Accessed 28 Dec. 2023.

Our Milky Way's Just-Right Spot in the Universe

For the third sweet spot, we find our Milky Way Galaxy in a unique place in the universe. Smethurst calls the area in which our galaxy resides "a void." She explains the universe can be thought of as a web of galaxies. The empty spaces within the web can be thought of as pockets of void, like holes in a sponge. Perhaps this void has protected our Sun from being disturbed. Nothing has flung us into a danger zone where radiation prohibits life.[27]

An international team of astronomers agrees we should be careful of "Copernican bias." In other words, we should be careful of the assumption that our station in the universe is commonplace. The scientific community has leaned heavily on that assumption since Copernicus demoted Earth from the center of everything 500 years ago.[28] But since the demotion, we have discovered we are indeed in a unique place in the universe—a pocket of void just right for the stable environment needed for life.

So here we are—in the perfect spot of the perfect solar system, which resides in the perfect spot of the Milky Way, which resides in the perfect spot of the universe. These three spots are not the only requirements for life like ours. Many more conditions must be met.

Smethurst estimates at least 100 sextillion stars exist, and perhaps 100,000 planets might be in a special spot to develop life. Given those numbers, that's a one in quintillion chance or 1 out of 1,000,000,000,000,000,000 chance we're here randomly.[29]

My husband plays the Texas Lottery every week. I love this about him—if he doesn't play, we have no chance of winning the

---

27 Scott, Joe (00:23:00-00:38:20). "Dr. Becky Smethurst On Being an Astrophysicist, English Accents, and the Cosmological Crisis" *Youtube,* uploaded by Joe Scott, 30 July 2020, www.youtube.com/watch?v=bxswbD4kSn4. Accessed 22 November 2023.

28 Kahlon, Gurjeet. "Milky Way Found to Be More Unique than Previously Thought." *The Royal Astronomical Society,* 20 Jan. 2023, ras.ac.uk/news-and-press/news/milky-way-found-be-more-unique-previously-thought. Accessed 23 Nov. 2023.

29 Smethurst, Becky. "Aliens Probably Exist." *Space at the Speed of Light: The History of 14 Billion Years for People Short on Time,* Ten Speed Press, 2020, p. 96.

millions. However, because of him, we have a 1 in 25,827,165 chance to win the Texas Lottery each week. He's been buying one ticket weekly for a couple of decades. We have yet to hit the jackpot.

So...What's the Point?

The point is you are more special than our special place in the universe. In July 2022, I eagerly awaited the media to publish photos from the James Webb Space Telescope. My favorite of these, and perhaps the most famous, is the photo of the Carina Nebula, a star-forming region in our very own Milky Way Galaxy showing an ethereal landscape of what looks like a massive dust storm in the sky with twinkling stars galore pouring forth some speech of good news, joy, and cheer all backdropped by my favorite color, midnight blue.[30]

Then, I saw a second photo from a coworker announcing the birth of his baby. A brand-new baby, so new little bits of birthing custard could be seen on his cheeks. With a perfect little face and eyes open, peering directly at the camera, I found myself in total awe. He was eternity bundled in flesh and a tiny hospital blanket, white with blue and red stripes. He is a bit of stardust, bearing the very image of God, alive with little seeing eyes that seem to tell me, "I am wisdom, humor, purpose, and love." The gorgeous Carina Nebula can't compare with stardust freshly bearing the image of God—although the portrait it paints does seem a thing worthy of this baby's gaze.

Some may declare we humans are too little to matter, less than a speck, yet our small blue marble, one thousand three-hundredth of Jupiter, holds life loved infinitely deep. So, it must not be

30 "NASA's Webb Reveals Cosmic Cliffs, Glittering Landscape of Star Birth." *NASA,* NASA, 12 July 2023, www.nasa.gov/image-article/nasas-webb-reveals-cosmic-cliffs-glittering-landscape-of-star-birth/. Accessed 23 Nov. 2023.

the size of matter that matters most. And if our smallness does translate to insignificance in the grand scheme of the cosmos, how can anything we say about the universe matter?

Conservationists create special places for the baby sea turtles they love, like a protected beach where cars cannot crush them and a square made out of flags and PVC pipes to keep the seagulls from carrying them away. When we look at the universe, we find ourselves in a protected place. It's reasonable for our eyes to widen in wonder and ask, *Could this mean, even though we are so small, we really are special in the grand scheme of things?*

It must not be the size of matter that matters most.

You are special. You're special to me; the reason why I labor over this book for you is to point you to the most beautiful person I know, Jesus. In the coming chapters, we'll come to know him better through what may be the most stunning and glorious light show known to humankind: the Total Solar Eclipse from the path of Totality.

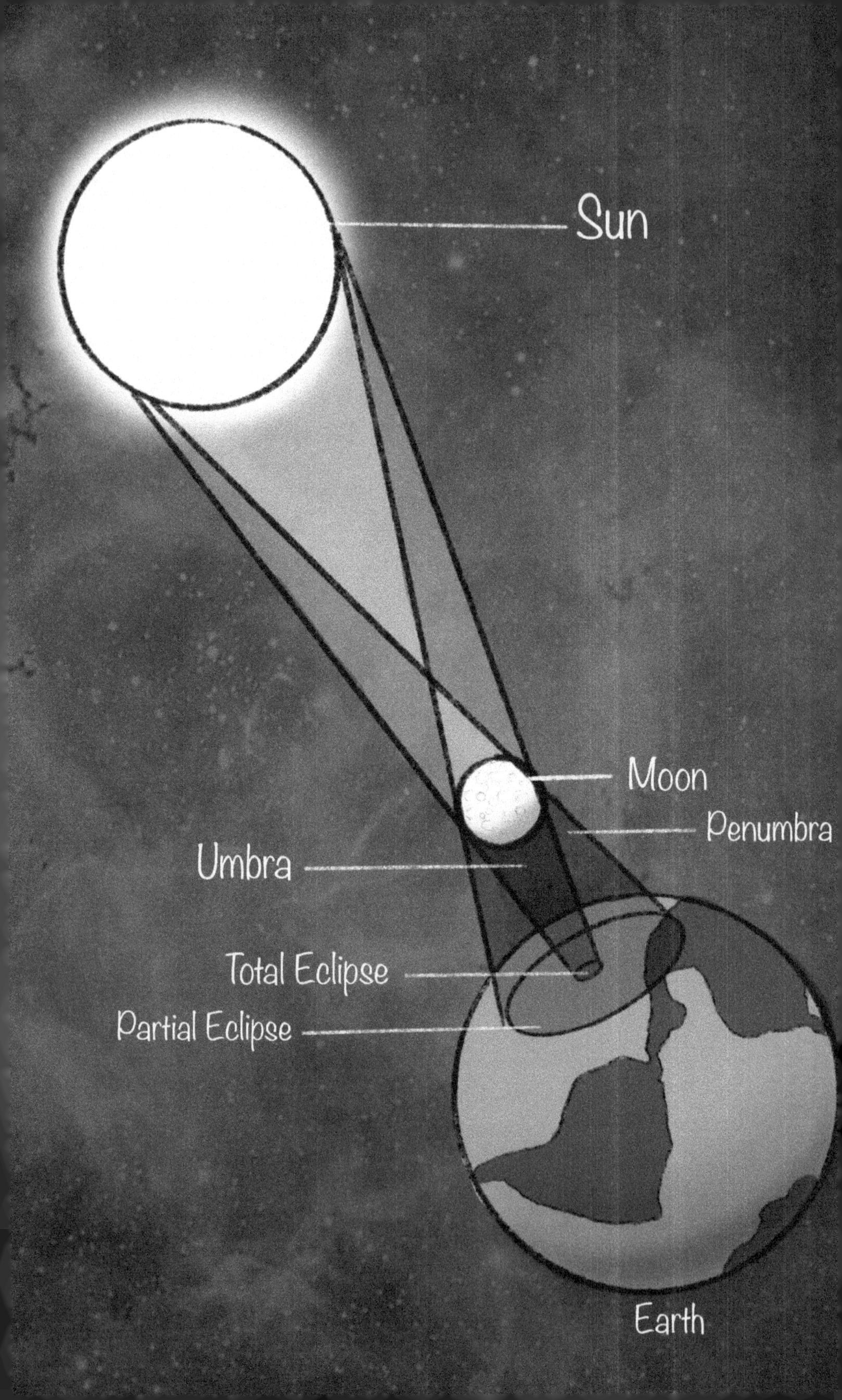

Sun
Moon
Penumbra
Umbra
Total Eclipse
Partial Eclipse
Earth

CHAPTER SIX

PRECISE MECHANICS OF THE ECLIPSE

The Great Designer

"Don't look!" Glory quickly covered something she was knitting when I entered her room. I can see it's baby blue, and I guess it's a blanket. She's almost always making something. I will have to wait to find out what this blue thing will become.

For Christmas 2022, she knitted me booties with sheep faces on top, complete with ears, eyes, nose, and mouth. They're cuffed and have an extra layer of padding in the sole. I wore them incessantly until 100-degree weather came in June. She made slippers for her future mother-in-law with Baby Yoda on top, a stuffed cat for her future brother-in-law that looks exactly like his beloved pet Chonkers, and a Santa kitchen towel holder for her future grandmother-in-law, who decorates her house in all things Santa Claus each Christmas.

Last week, her fiancé planned to take her on a fancy date, so she made him the cutest, tiny needle-felted bear, his favorite animal, to give him when he picked her up.

"Do you want to see what I was knitting? It's finished!" Glory smiled.

"Yes, of course!" I answer.

She laid out a light blue baby blanket with white edges on the floor. A large gray whale shooting water from its blowhole smiles at me from the middle. A pink, knitted heart dangles from the top right corner of the blanket.

"Oh my goodness! Did you design it yourself?" I asked.

"Yes, I drew it out on a grid first."

"Amazing! I assume it is for your baby?"

"Yes, Mother, I can't wait!"

I love it when she calls me Mother. Glory is not expecting a baby, but she's been dreaming about becoming a mother since third grade. I write this in the summer of 2023, a year before she and her fiancé plan to marry. Then, he plans to attend medical school, and she hopes to have a baby as soon as he finishes.

Glory designs and creates for her fiancé and his family because she's in love. She's even in love with her future children, who are currently only a twinkle in her eye and the deep desire of her heart. Could it be we have a God who loves like Glory? Or even more than Glory? One who loves us so much that he designed and created Earth and space to meet our desires and needs when we were only a dream in his heart?

Design or Chance?

My friend Ann pointed out that if you had a million monkeys typing on a million typewriters for a million years, they would never ever end up by chance typing even the first paragraph of the Declaration of Independence. To believe the universe was created by chance is to hold a belief that, statistically speaking, is similar.

Is it possible the Total Solar Eclipse was designed on purpose far in advance of our arrival to show us a deep sense of love? It's a big idea and difficult to imagine a love like that, but it makes more sense than assuming it happened by accident. The Total Solar Eclipse has an intricate design behind it. When we experience a Total Solar Eclipse from the path of Totality, we may feel small in the cosmos and yet also compelled, even called to something. When we look outside ourselves through prayer, we find help is there. When we study our place in the universe, we find our spot so special that it's strange to think it came about by chance.

Is it possible the Total Solar Eclipse was designed on purpose far in advance of our arrival to show us a deep sense of love?

Precise Mechanics of the Perfect Eclipse

A perfect Total Solar Eclipse requires three bodies correctly aligned: the Sun (the light source), the Moon (the eclipsing body), and an observer platform (the Earth upon which we stand).

What do we mean by a "perfect" Total Solar Eclipse? A perfect Total Solar Eclipse is one in which the Moon eclipses the Sun in such a way as to block the photosphere of the Sun (the brightest, visible layer) yet allow the Sun's chromosphere (a layer of plasma on top of the photosphere) and corona (the Sun's outer atmosphere) to remain visible.[1]

Perfect ratios involving the size and position of the Sun, Moon, and Earth produce this perfect eclipse. The Sun is 400

1 Gonzalez, Guillermo, and Jay W. Richards. *The Privileged Planet: How Our Place in the Cosmos Is Designed for Discovery.* Regnery Publishing Inc., 2004. pp. 7-9.

times larger than the Moon. Additionally, the Sun is 400 times farther from Earth than the Moon is.[2] This wordy explanation can also be expressed with the following mathematical equations, where the radii and distances used are only rough estimates based on NASA data.[3,4]

Equation 1

$$\begin{aligned}&\textit{Radius of Sun} \div \textit{Radius of Moon}\\&\simeq 433{,}288 \text{ miles} \div 1080 \text{ miles}\\&\simeq 400.\end{aligned}$$

Equation 2

$$\begin{aligned}&\textit{Sun's Distance from Earth} \div \textit{Moon's Distance from Earth}\\&\simeq 93{,}000{,}000 \text{ miles} \div 232{,}500 \text{ miles}\\&\simeq 400.\end{aligned}$$

This pure geometry explains mathematically the situation that provides a perfect eclipse.

Not a numbers person? The simple science experiment on the next page can help you visually understand the precise positioning required for a Total Solar Eclipse.

2 "Why Do Eclipses Happen? - NASA Science." NASA, NASA, Sept. 2023, science.nasa.gov/eclipses/geometry/. Accessed 23 Nov 2023.

3 Williams, David R. "Sun Fact Sheet." *NASA,* NASA, 17 Nov. 2022, nssdc.gsfc.nasa.gov/planetary/factsheet/sunfact.html. Accessed 23 Nov 2023.

4 Williams, David R. "Moon Fact Sheet." *NASA,* NASA, 20 Dec. 2021, nssdc.gsfc.nasa.gov/planetary/factsheet/moonfact.html Accessed 23 Nov 2023.

SCIENCE EXPERIMENT #2

Create the Perfect Total Solar Eclipse

Materials

- Friend
- Dinner plate
- Dessert plate, smaller than the dinner plate

Procedure

1. Have your friend hold a dinner plate at approximately the same height as your eyes, where the top of the dinner plate is facing you.
2. Hold a dessert plate at arm's length where the top of the plate is facing you, and it is centered over the front of your friend's dinner plate. The plates should be touching. Notice the dessert plate does not cover the dinner plate. You can see the surface of the dinner plate surrounding the dessert plate.
3. Slowly take steps backward, keeping the dessert plate perfectly aligned with the center of the dinner plate. Notice the more steps you take, more of the dinner plate is covered by the dessert plate.
4. Continue to carefully take steps backward, keeping the dessert plate perfectly aligned with the center of the dinner plate until only a very thin edge of the dinner plate can be seen. At this position, the dessert plate and dinner plate appear to be the same size.
5. Continue to take steps backward. It may become difficult to keep the dessert plate perfectly aligned

with the center of the dinner plate because you can no longer see the dinner plate at all. The further away you are, the smaller the dinner plate appears. The dessert plate covers the edge of the dinner plate and the space around it.

Conclusion

It is only when the dessert plate is precisely positioned that it appears the same size as the dinner plate, and a "perfect eclipse" occurs. It cannot be too close or too far away from the dinner plate, and the centers must remain aligned. So it is with our Moon, Earth, and Sun during a Total Solar Eclipse.

This perfect positioning and sizing of Earth, Moon, and Sun not only lend themselves to a Total Solar Eclipse, but they also lend themselves to the conditions necessary for life, as we saw in the previous chapter. No other eclipse in the solar system is like it.

Rarity of the Perfect Eclipse

On her twenty-second orbit around the gas giant that is Jupiter, the Juno spacecraft captured a picture of the shadow of one of Jupiter's moons, Io, during a solar eclipse. When Io eclipsed the Sun, its shadow crawled across Jupiter like a dark, ominous black hole.[5] This looks nothing like the shadow of our Moon traveling across Earth. Our shadow looks like a fuzzy gray spot.[6,7] If an

5 "Junocam: Image Processing." *Mission Juno,* Southwest Research Institute, 23 Dec. 2022, www.missionjuno.swri.edu/junocam/processing?phases%5B%5D=PERIJOVE%2B22. Accessed 01 Jan. 2024.

6 "Moon's Shadow on Earth During Solar Eclipse." *NASA,* NASA, 8 Apr. 2014, www.nasa.gov/image-article/moons-shadow-earth-during-solar-eclipse/. Accessed 1 Jan. 2024.

7 Godbole, Ravindra. "Umbra and Penumbra - Lunar and Solar Eclipse." *YouTube,* YouTube, 22 Dec. 2019, www.youtube.com/watch?v=olSaCmUztJw. Accessed 30 Dec. 2023. (See this science experiment to better understand what makes the shadows look differently).

observer could stand on Jupiter and live to tell the tale of the Io/Jupiter eclipse, they would not describe a diamond ring or corona because all light from the Sun, including its chromosphere, would be completely blacked out.

On Jupiter, that observer would miss the deeper truths we glean from our Total Solar Eclipse. On Jupiter, that observer would miss the glory we're given through ours.

Our solar system consists of eight planets, nine if you grew up in the 1980s, and thus inherently demand Pluto remain in the count. As I write this, we know of 290 traditional moons. Mercury and Venus are the only planets with no moons. Earth is the only planet with one. Jupiter has 95 moons, and Saturn has 146.[8]

Out of all the possible eclipses, ours is the only one that is perfect. And ours is also the only one with observers. The mechanics required for a Total Solar Eclipse overlap the requirements for life on Earth.[9]

Think about that. The only perfect eclipse in our solar system is the very one we humans have the privilege to experience. We can't help but recognize the beauty of its design.

Recognizing Design

What have we experienced that comes together in form and function without a desire and a designer at the root of it?

I write software for NASA and European Space Agency (ESA) space missions. That may sound glamorous to some people, but what it really means is that I spend an inordinate amount of my time determining what symbols I should type in an intentional

8 NASA. "Moons" *Solar System Exploration: Our Galactic Neighborhood* Web Site, 1 June 2023, https://solarsystem.nasa.gov/Moons/overview. Accessed 1 June 2023.

9 Gonzalez, Guillermo, and Jay W. Richards. *The Privileged Planet: How Our Place in the Cosmos Is Designed for Discovery*. Regnery Publishing Inc., 2004. p. 4-7, 10.

way to achieve a desired result, an application that meets a vast array of requirements.

Suppose I pressed the keys on my keyboard randomly during the work day. In that case, there is a 0% chance any sensible or even barely functioning application would appear. We could never take such a random approach and expect to please NASA or the ESA. No one needs to be a computer scientist to know this.

Wisely, NASA and the ESA assign the work to a skilled designer who can meet the form and function specifications and achieve the desired outcome. The same applies to cooking.

What would cooking be like without the design of a recipe?

Blindfold a stranger. Invite her to your pantry to choose the first three items she touches. Guide her to your refrigerator, open the door, and let her choose the first two items her fingers brush against. Now, put these five ingredients in a pot. Spin a wheel to select a temperature setting. Roll some dice to choose the cooking time. We need no trained scientist, mathematician, or statistician to know the odds are low that this meal will bring pleasure.

From coding and cooking to composing, it's easy to recognize and appreciate a good design.

But invite a trained chef. Invite her into your house with her eyes open to design and create a meal for you, and the odds are high that the meal will delight you.

Music is also made by design.

Allow a small child who has had no piano lessons to bang on the keys. Although we may take some pleasure in his delight of finding the power to make a cacophony of noise, we do not expect the resulting sound to please us. But bring in a pianist who plays a song composed, and our ears are delighted.

From coding and cooking to composing, it's easy to recognize and appreciate a good design. No pleasing software, meal, or music exists without some planning and expertise behind it. It's reasonable to believe that designs discovered in nature are also rooted in planning and expertise.

Design from Discovery

Scientists are discoverers. They unearth the designs behind the world around us, uncovering biological and astronomical wonders. Engineers use those scientific discoveries as blueprints to create fantastic inventions for humans. Perhaps the most famous example of mimicking biology is VELCRO®. Digital flight data recorders (DFDRs) and swarm robotics also find their inspiration from nature.

Scientists unearth the designs behind the world around us, uncovering biological and astronomical wonders.

VELCRO®

One of the most well-known examples of biomimicry, copying designs found in nature, is VELCRO.

After an afternoon of hunting in 1941, George de Mestral, a Swiss electrical engineer, came home with cockleburs embedded in his pants and his dog's fur. Fascinated by how well they clung, he studied their design under a microscope and realized burs are designed with hundreds of tiny hooks that grab and hold on tightly to anything with loops.

Inspired by these burs, de Mestrel recognized hook and loop applications to replace fasteners. Still, it took him 10 years to successfully mimic this impressive design, invent VELCRO, and file for patents in various countries. Ten years later, VELCRO became a significant contributor to the Apollo program.[10]

Covering approximately 275 square feet of the Apollo 11 spacecraft and its accessories, VELCRO Brand hook and loop secured instruments to space suits, anchored astronauts' feet in their boots, secured camera equipment, and served many other purposes. The genius design of VELCRO provided the "durability, reliability, strength, and versatility" needed to solve astronauts' challenging problems navigating a zero gravity environment in bulky space suits that limited their dexterity.[11]

George de Mestral's brilliance isn't defined by creating his own original design. His genius is defined by recognizing the existing intelligent design of burs and reproducing that design in new applications for the betterment of humanity.

VELCRO is one of many examples of bioinspiration, the development of novel creations inspired by a great design found in nature. Another example includes digital flight data recorders designed after the anatomy of a bird.

Digital Flight Data Recorders

In 2004, I led a small team to write a computer program for the United States Air Force, which presented data from a digital flight data recorder (DFDR), an airplane's black box, in a human-readable format. A black box records information from its airplane's flight, such as speed, velocity, heading, etc. If the aircraft crashes, investigators retrieve the black box from the crash site and then use its data to try to understand what caused the accident.

For this to work, a black box must meet at least two criteria: (1) it should be easy to find among the chaos of the wreckage, and

10 "History of Velcro Companies - Our Timeline of Innovation." VELCRO®, VELCRO®, 14 Mar. 2023, www.velcro.com/original-thinking/our-timeline-of-innovation. Accessed 24 Nov. 2023.

11 "Securing Success for NASA Astronauts - Velcro Companies." *VELCRO®*, VELCRO®, 20 Feb. 2023, www.velcro.com/original-thinking/securing-success-for-nasa-astronauts/. Accessed 24 Nov. 2023.

(2) it must somehow survive the blow of impact. To meet the first requirement, designers chose to paint "black" boxes bright orange. To withstand the impact of a plane crash, which may experience 3,400 Gs (3,400 times the force of gravity), black boxes are designed after a woodpecker's head.

A woodpecker's head is designed with a hard beak, spongy skull bones, a thin layer of fluid around the brain, and a hyoid layer, which can be thought of as a very long tongue that wraps around the brain and acts as a protective cushion. A woodpecker's beak experiences repeated hits of 1,200 Gs each time it pounds a tree, yet its brain remains undamaged. In 2011, Engineers from the University of California, Berkeley, mimicked this design to create a more powerful DFDR that can withstand 60,000 Gs.[12]

Engineers continue to mimic designs in nature.

It's reasonable to respect the engineers who created such a strong DFDR. It's even more logical to be in awe of the one who designed and created the woodpecker's beak. Engineers continue to mimic designs in nature. Some designs are so complex that we're still learning how to copy them, as seen in swarm robotics.

Swarm Robotics

My friend Paul contributed to the development and testing of an instrument called the MASPEX, which stands for "MAss Spectrometer for Planetary EXploration/Europa." The MASPEX is one in a suite of nine instruments that will fly on the Europa Clipper spacecraft, currently scheduled to launch in late 2024 to study Jupiter's moon, Europa.

Paul described the MASPEX to me as a dog's nose, which can sniff up the atmosphere, separate its molecules, and identify its

12 Marks, Paul. "Woodpecker's Head Inspires Shock Absorbers." *New Scientist,* New Scientist, 4 Feb. 2011, www.newscientist.com/article/dn20088-woodpeckers-head-inspires-shock-absorbers/. Accessed 24 Nov. 2023.

chemical elements and compounds, such as water vapor, carbon dioxide, and molecular oxygen. My friend Rebecca, who leads the team that will command the instrument when it's flying through space, told me we hope to fly this dog's nose through one of Europa's geysers, plumes we observed through NASA's Hubble Space Telescope. Scientists believe these plumes are ejected from a subsurface ocean underneath the icy crust of Europa. Astonished, I asked her, "Really? We really think we can fly through a geyser?"

"Really," she grinned.

Ethan Schaler, a robotics mechanical engineer at NASA's Jet Propulsion Laboratory in Southern California, dreams of dropping a swarm of robots to collect data about other parts of our universe, like Europa's oceans, where it's unsafe for human life to explore.[13]

Perhaps if the MASPEX is able to fly through a geyser and detect signs of life, another spacecraft can be sent to drop a swarm of tiny robots that can break through the moon's icy crust, swim together through the subsurface ocean, and tell us more. For now, it's only a dream.

Engineers look to swarms found in nature, such as huge groups of starlings swooping and swirling together in the sky or schools of fish swimming in the same direction like one organism.[14] We haven't figured out how to create a robotic swarm as large and organized as those found in nature yet, but engineers continue to try.

From VELCRO to a swarm of robots, engineers look to the intelligent, impressive designs found in nature for inspiration. We spend extreme effort to mimic and repurpose God's original designs to attempt to serve our desires, needs, and curiosity.

13 Pamer, Melissa. "Swarm of Tiny Swimming Robots Could Look for Life on Distant Worlds." *NASA Jet Propulsion Laboratory,* NASA, 28 June 2022, www.jpl.nasa.gov/news/swarm-of-tiny-swimming-robots-could-look-for-life-on-distant-worlds. Accessed 24 Nov. 2023.

14 "Swarm Robotics." *PBS LearningMedia,* NOVA, 11 Jan. 2021, klrn.pbslearningmedia.org/resource/nvmms.sci.eng.swarm/swarm-robotics/. Accessed 24 Nov. 2023.

So...What's the Point?

The point is that it's reasonable to assume a great designer planned everything in the universe, including the Total Solar Eclipse, one of the most glorious light shows in existence. This light show requires precise positioning of the observer, Earth, Sun, and Moon. Its raw beauty surpasses that of any choreographed fireworks show.

Still, much literature describes it as a random accident. In the TIME Special Edition Beautiful Phenomena, senior science writer Jeffrey Kluger states with confidence, "[T]he Moon was not placed in space for our entertainment. In fact, it was placed there by accident." He goes on to imply it's sinful to view the Total Solar Eclipse as more than an accident by using the word temptation in the following sentence: "Even for scientists, there can be a temptation to see the eclipse as something intended to thrill, a sky show put on for the only species in the solar system able to appreciate it."

Why is it wrong for a scientist, or any human, to view the eclipse as a purposeful light show? Given much of our experience in life, from coding and cooking to composing, does it not seem a reasonable hypothesis that it was created by a designer with intention, care, planning, and purpose?

Does it not seem a reasonable hypothesis that it was created by a designer with intention, care, planning, and purpose?

No one has the audacity to say that VELCRO, DFDRs, or robotic swarms were created by chance, yet some confidently state that cockleburs, woodpeckers, and starlings, the very blueprint for each invention, were formed by accident.

Will you assume with me in the coming chapters that the

eclipse was designed on purpose long before we came to be on Earth? If we are courageous enough to dream of flying mass spectrometers through geysers in outer space in hopes of finding fantastic discoveries, then surely we can have the courage to dream of seeking and finding the creator of everything. In my search, I have found the faith and evidence to believe the love of the creator is much greater than Glory's love for her future husband and children still yet to be conceived.

CHAPTER SEVEN

CREATED WITH PLEASURE

God's Commands and Desires

When we moved into our first home, a little red brick one-story house directly across the street from the playground of a public elementary school, Jon and I worked full-time at the same company writing software for the C-5 Galaxy, the largest cargo plane in the United States Air Force.

The company had an onsite daycare, which was critical to our survival since we had two little girls: Joy, age four, and Glory, age two. Before we bought the little house, we lived in an apartment. When we parked and walked inside, we seldom saw a neighbor. That changed when we moved into the house in 2003. The neighborhood was full of young children and mothers who did not work outside the home.

"Oh, you will have to bring six dozen cookies to Laurie's Christmas cookie exchange in December!" Lisa, the previous owner of the house, said when we bought it.

Cookies? Doesn't she know I have to work? I don't have time to bake cookies! "Oh, okay," I said.

Five months later, in December, I managed to bake the cookies for the cookie exchange, squeezing it in on my day off.

Courtney and Caitlin, two neighborhood girls the same age as ours, rode up to our van on their three-wheeler bikes the minute we parked in our driveway at the end of most workdays.

"Hi! Can Joy and Glory play?" They beamed with excitement as soon as we opened the door to get out of the van.

Are you kidding me? My girls have been playing with other kids all day at daycare. We're exhausted from a long work day and need to get inside and recharge.

"No, not today," I smiled politely while I lifted Glory out of her car seat and hustled my girls to the front door. "We need to go inside and eat dinner."

Life was stressful. Jon and I argued a lot. Our marriage was not a happy, joyful one, and I wondered if we would become a divorce statistic. Two years after the move and six months after our third child, Freedom, entered the world, I felt God tell me, *Go home and focus on loving your family.*

I have never heard God audibly, but sometimes I hear his voice speak a mandate to my mind, like a gift. It feels like it comes from outside myself. It doesn't sound like me, and it doesn't feel like my idea. The mandate usually surprises me and causes respect to rise inside me, enabling me to obey it.

I never desired to be a stay-at-home mom, and my situation could not have been more convenient for a mother working outside of the home. Jon and I drove in the same car to work at the same place where our children attended daycare, and we visited them during breaks and at lunch. Why would I become a stay-at-home mom and let go of the benefit of two incomes and the work I enjoyed?

I did it because of the voice. The words "Go home and love

your family" were more than a command; they caused something inside me to change. Sometimes, God speaks like that. He speaks over us with a creation power. Perhaps I can explain by looking at one of God's most repeated mandates, "Be strong and courageous. Do not be afraid."[1]

Commanding, Encouraging, Creation Words

We read the words, "Be strong and courageous. Do not be afraid." many times in the Bible. For most of my life, I understood it as a command I could not conceivably obey.

In my late forties, my friend Jill taught me to see God's imperative in a new light, not as a command, but as an encouragement.[2] The words came like the voice of a father who supports his daughter in an upcoming conquest with four special words tacked on the end, "Don't be afraid. Be strong and courageous, for *I am with you.*" This encouragement empowers us to move forward in confidence.

"Don't be afraid. Be strong and courageous, for I am with you."

A year or so after finding encouragement in the command, I came to know the words *Don't be afraid. Be strong and courageous* in an even newer light. Lying on the hardwood floor one day, stretching my legs while listening to the Bible, I heard those familiar words in Joshua 1:9, "Fear not! Be strong and courageous."

1 The words appear directly in the NIV in the following places: Deuteronomy 31:6, Joshua 1:7 Joshua 1:9, Joshua 10:25, 1 Chronicles 22:13, 1 Chronicles 28:20, and 2 Chronicles 32:7. The sentiment appears in many places including but not limited to Matthew 10:28, Luke 12:4, Luke 12:7, Luke 12:32, and 2 Timothy 1:7.

2 McCormick, Jill E. producer and show host. "BONUS #2: Grace in the face of anxiety, fear, uncertainty, and worry." Grace in Real Life Podcast, 23 March 2020. https://www.jillemccormick.com/bonus-2-grace-in-the-face-of-anxiety-fear-uncertainty-and-worry/. Accessed 24 Nov. 2023.

This time, they sounded different and felt like electricity. They reminded me of creation words, the words God spoke in the beginning when God said, "Let there be light," and light was; when God said, "Let dry ground appear," and it did. God told Joshua, "Be strong and courageous," and Joshua was.

God says words like that, not only as commandments to obey, not only as an encouragement to inject courage, but also as creation words that change us in a moment.

God says words like that, not only as commandments to obey...but also as creation words that change us in a moment.

That's how I became a stay-at-home mom. I didn't plan to become one. God spoke a creation word over me. Even though I had a job that seemed so perfectly suited to my skills and situated in the best possible way for our family to thrive, I suddenly had the courage to step away from one role and into another, and I became someone other.

God spoke, and a stay-at-home mom was.

The Most Famous Creation Story

Consider the most famous creation story found at the beginning of the Bible: When God spoke, and everything was. The first two sentences remind me of what Becky Smethurst taught us about our special place in the universe, that pocket of void that appears to keep our system stabilized and undisturbed. Genesis 1:1-2 says, "In the beginning, God created the heavens and the earth. The earth was without form and *void;* and darkness was on the face of the deep" (ESV, emphasis added).

God spoke creation into existence: light, sky, water, land, vegetation, the Sun, the Moon, and the stars. He spoke swimming, flying, and land creatures into existence. It was all his idea. He spoke out of his heart, mind, will, desires, and emotions. There's a saying, "Out of the abundance of the heart, the mouth speaks" (Matthew 12:34-37). Out of the abundance of the heart of God, his mouth spoke perfect creation into existence. And then God involved his hands.

He formed us out of the clay of the ground and breathed life into our earthen bodies. If out of the abundance of the heart, the mouth speaks, how much more so do the hands of an artist build, work, and create? With his hands, God planted the Garden of Eden and sculpted Adam from the clay of the ground. Every work of art, from garden to sculpture, is birthed by an artist and shows us something inside that artist.

Looking Inside God's Heart

The Garden of Eden shows us God's heart. This work of art, this manifestation of his heart, was more than a painting humans could see with their eyes. He wanted us to not just see his heart but also to *experience* his heart, to *know* his heart through a variety of senses.

When we walk through a garden, we can see it, and we can also hear it. We can hear it in the birds, in the unseen wind dancing with the leaves of trees, and in the footsteps of a friend. We can feel a garden, its dirt pressing against our feet, the soft petals of a rose at our fingertips. We can smell it, the fragrance of honeysuckle. We can taste a garden in the fruit of its vine. We can see it, hear it, touch it, smell it, taste it, and know the Creator's heart is good. And good is pleasurable.

Eden Is Pleasure

The Hebrew word for pleasure is *'êden.*[3] The first time the word *'êden* was used in the Bible with a little *e* is when Abraham's wife Sarah heard she would have her first and only child in her old age. She laughed to herself as she wondered: *After I am worn out, and my husband is old, will I now have this pleasure ['êden]?"* (Genesis 18:12-15)

Knowing this, we can re-read the book of beginnings and replace the name of the garden, Eden, with the word Pleasure. Read this passage, where I've taken the liberty of swapping in the translated word:

> Now the Lord God had planted a garden in Pleasure and there, he put the man he had formed. The Lord God made all kinds of trees grow out of the ground—trees that were pleasing to the eye and good for food. In the middle of the garden were the tree of life and the tree of the knowledge of good and evil. A river watering the garden flowed from Pleasure; from there it separated into four headwaters… which wound through the entire land in each direction. (Genesis 2:8-10)

God "planted a garden in Pleasure" and "a river watering the garden flowed from Pleasure." God created the Garden of Eden not for business or out of obligation but for pleasure with pleasure. We can explore and know God's desires by studying all he offered in the beginning.

God's Desire to Share His Likeness

God desired to share more with us than just his heart manifested in a garden. He desired us to be the unique image of his beauty in creation.

3 "H5731 - 'êden - Strong's Hebrew Lexicon (NIV)." Blue Letter Bible, Mar. 1996, www.blueletterbible.org/lexicon/h5731/niv/wlc/0-1/. Accessed 24 Nov. 2023.

Genesis 1:26 says, "And God said, 'Let us make man in our image, after our likeness.'" We are made to put the beauty of God on display. God is not arrogant. He's not a god who separates himself far above his people, looks down, and says, "You little ugly ones, look up and see how beautiful I am!" No, he is a God who desires to share his beauty with us, to be close to us in every way, even in likeness.

God's Desires Are Beautiful and Pure

We hope the gifts we receive are given out of pleasure. No one wants a gift given out of obligation. No one wants a gift given in grumbling.

We live within walking distance of one of my favorite fast-food restaurants, Chick-fil-A. One of the things I love about Chick-fil-A is the way the workers respond when I say, "Thank you." They're trained to say with a smile, "It's my pleasure!" I don't know if the people at Chick-fil-A always mean it from the heart when they say, "It's my pleasure"—I know it's part of their job, and they're getting paid to say it, but it seems sincere.

No one wants a gift given out of obligation.

We don't have to question God's sincerity. No one is paying God. We know everything God does is because he wants to do it—that's part of what makes one a god, isn't it? The ability to do whatever you wish because you have all the freedom and all the power to do it?

One thing that makes our God beautiful is that his desires are beautiful. God is not some strange, lazy, self-indulgent god who lounges on a sofa while half-naked servants entertain him, fan him with big feathers, and feed him grapes. No, that's not what gives him pleasure. We have a God whose desires are pure, noble, right, lovely, and beautiful in every way. He is a God who desires pure, lasting, beautiful pleasure and wants to share it with us.

God's Desire for Togetherness

Genesis 2:7 says, "Then the Lord God formed a man from the dust of the ground and breathed into his nostrils the breath of life, and the man became a living being."

I love to imagine the beauty of the closeness of Adam and God in this verse. Adam, his every part, sculpted by God's very hands from the clay of the Earth. And then God gets so very close—nose to nose, mouth to mouth, and breathes the breath of life into Adam. Adam's lungs fill. I imagine with his very first exhale, he opens his eyes, and the first thing he sees is God's face close to his. The first thing he smells is God's sweet breath. The first thing he feels is the sweetness of God's presence.

> We are made by God through his purpose and by his pleasure so that we, as observers on this Earth, can enjoy each other and all of creation, which was also made through his purpose and by his pleasure.

When I think of this moment, I have always imagined the wonderment in Adam's face. One day, when contemplating the moment, I heard God say, "Look at the wonderment on *my* face." I heard God's response to Adam's eyes come alive, "Wow, a son in my likeness, breath of my breath, my desire met." God created the son he wanted. What can match the face of a good father holding his baby for the first time? The delight, the joy, the pleasure is so great.

God's Desire to Share Even More

It wasn't enough for Adam to be this close to God. God wanted *more* for Adam. God desired for Adam to feel the same wonderment God felt. In Genesis 2:18, God said, "It isn't good for the man to be alone. I will make a helper suitable for him."

God caused Adam to fall into a deep sleep, took a rib from Adam, and made a woman from it. God brought her to Adam. Now, if anything can match the marvel on a father's face when he holds his baby for the first time, it is the awe on the face of a groom when he beholds his beloved bride on their wedding day.

When God brought Eve to Adam, I wonder if Adam recognized the marvel he felt as the marvel he saw on the face of God when Adam first opened his eyes. I wonder if Adam mimicked his Father when he said, "Wow, bone of my bones and flesh of my flesh."

Adam's desire was met. Genesis 2:24 says, "That is why a man leaves his father and mother and is united to his wife, and they become one flesh."

One flesh. We were made for a togetherness so great that it is called oneness. Marriage reflects the intimacy and oneness God longs to have with us. It reflects the intimacy God planned for us.

We are made by God through his purpose and by his pleasure so that we, as observers on this Earth, can enjoy each other and all of creation, which was also made through his purpose and by his pleasure. He wanted us here, in this special spot of his creation, so we can experience his goodness.

So...What's the Point?

The point is, according to the most widely read, most treasured, most propagated story of beginnings, you and I were made on purpose with pleasure because we were wanted. In the beginning, everything was good and perfect, but as we'll see in the next chapter, evil attempted to pervert God's perfect creation, which reminds me of the beginning of a Total Solar Eclipse.

Everything is right with the Sun and Earth before a Total Solar Eclipse crosses our path. The Sun shines perfectly, as it should, giving us daily light so we can see, and giving energy to grow plants to provide the necessities of life: food and oxygen. But then, something strange happens. Darkness seems to separate us from our necessary light, causing some people to feel a great sense of dread when they experience it. But as we'll see in the coming chapters, no darkness can eclipse God's good light, plan, and purpose for your life. In the beginning, God blessed Adam, Eve, and all their offspring with beauty and freedom. No darkness or curse can eclipse or overtake that blessing.

No darkness can eclipse God's good light, plan, and purpose for your life.

Darkness tries to overtake our lives. I thought it might take over my marriage, but God stepped in and helped us. Becoming a stay-at-home mom in a little house was not my idea or plan, but God used this transition to pave the way for our family to find fellowship with God's people and to come to know Jesus on a deeper level. God shaped me into the role of a homemaker and sustained Jon as a provider, but God was the true homemaker and provider. Our shortcomings could not eclipse his plan for us.

In the remaining chapters, we'll discover that no darkness can eclipse God's love for the world. We'll align God's love manifested in the story of the world with the design of the Total Solar Eclipse and its five phases:

1. First Contact (Growing Partial Eclipse): Separation from God and the Fall
2. Second Contact (Diamond Ring): The Old Covenant
3. Totality (Corona): Jesus on Earth Crucified, Buried, and Raised to Life
4. Third Contact (Diamond Ring): The New Covenant
5. Fourth Contact (Shrinking Partial Eclipse): Reconciliation with God and Restoration

Shall we turn the page and keep going? In the next chapter, I'll share the answer to a critical question that bothered me for years. *Why did God make a way for evil to enter the world?*

CHAPTER EIGHT

FIRST CONTACT

The Fall

"Don't do it!" Joy cried out. She and I were reading a new children's picture Bible at the kitchen table. It's the first time I remember reading the Bible with her. She was four years old, and we had just read about all the beautiful things God had made in the beginning.

The Sun, Moon, Earth, and stars.

The plants, fish, birds, and animals.

Adam and Eve, the garden with all its trees.

Creation was exquisitely illustrated in this book, and we soaked it all in.

We read about God's warning, "Don't eat from the fruit of the tree of knowledge of good and evil, for when you eat of it, you will surely die."

We turned to the page when a snake appeared in the forbidden tree and tempted Eve. Joy exclaimed, "Don't do it!" as if Eve could hear her voice. The concern on her face surprised me.

I sensed her appreciation and delight in all God created as we read. Given God's warning on the previous page, she understood what it meant if Adam and Eve ate the fruit.

We turned the page.

We didn't need to read a word.

The illustration spoke.

Eve partook.

And so did Adam.

Tears silently spilled down Joy's four-year-old cheeks.

I will never forget it.

People refer to Adam and Eve's choice to distrust God as the Fall. I see it symbolized in First Contact of the Total Solar Eclipse.

First Contact Phase of the Eclipse

The Total Solar Eclipse begins with First Contact—the moment when the Moon takes its first tiny bite out of the Sun, partially eclipsing our star, but only barely. You will not notice First Contact unless you put on protective paper glasses and look up at precisely the right time, a time you cannot know unless you understand the mathematics and heavenly clockwork involved. When I saw my first Total Solar Eclipse in 2017, I didn't study the math behind the eclipse, so I didn't know when it would begin. I was thankful my brother-in-law CJ sent us an email that said, "It looks like the partial phase starts at 11:56:44 AM (CDT)." We trusted that fully, so much so that that's precisely the time we decided to look up at the sun with our paper glasses. I didn't see the moon, but I did see a tiny piece of the sun blacked out, missing.

First Contact is a growing partial eclipse. In 2017, this phase lasted approximately one hour and thirty minutes. Little is noticed on Earth at first. It's a time when you look away from the sky and go about your business. You put on your safety spectacles every now and then to see the progression of the black disc. The tiny bite grows. The sun diminishes, slowly taking the shape of a crescent.

Even when the Sun was 50 percent covered, the day felt incredibly and normally bright, as though no piece of the Sun was darkened. This surprised me. I marveled at the Sun's power when covered even more, 70, 80, 90 percent. Even when the Moon masked a vast majority of the Sun, its heat and light felt unaffected, unmodified. That's how powerful the Sun is.

Even when darkness attempts to eclipse God and move between him and us, he still shines bright.

I think God is like this. Even when darkness attempts to eclipse God and move between him and us, he still shines bright, really brightly.

If it were not for the special glasses that allowed me to safely peer at the Sun and see its crescent shape, I would not have known anything unusual was happening. It reminds me how evil sneaks in slowly on a life. You may not notice its entrance or growth unless you are aware of its presence and are keeping watch.

Though I may not have known what was happening in the heavens without the forewarning given by those who study the stars, I may have noticed the crescents of light dancing on the ground.

Crescents of Light

During First (and Fourth) Contact, crescents of light delight us as they dance on the ground under the trees. What causes these?

In an open space, where no trees cast shadows on the ground, the ground is fully covered and flooded with sunlight. The leaves of trees, however, block sunlight, preventing some rays from touching the ground. It's in this space, under the trees, where crescents of light delight us. A true reflection of the eclipse overhead, you will notice the crescents on the ground are inverted from the crescent in the sky—if the crescent in the sky opens to your right, the crescents on the ground will open to your left. Science Experiment #3 demonstrates.

Create Your Own Shadows

If you're preparing to watch a Total Solar Eclipse in an open area with no trees, I don't want you to miss the crescents of lights! Bring an arrangement of fake flowers (to simulate tree leaves), a colander, or a piece of paper with holes punched in it. Using any of these items to cast shadows on the ground will reveal the crescents of light. If you don't have any of these items, put your hands together so your fingers overlap like a hashtag and view the crescent shadows in the spaces between your fingers.

The crescent shadows give us a clue something is happening in the sky. We can look up with special glasses and know a dark disc attempts to eclipse the Sun. Without paying attention to the hints or purposefully looking up with paper glasses, the majority of First Contact occurs without our knowledge. This phase symbolizes the Fall in the story of the world and also symbolizes the way evil attempts to stealthily take over our hearts.

First Contact and the Story of the World

The story of the world, as told in the Bible, can be directly mapped to the phases of the Total Solar Eclipse. Once you see it, it's hard to unsee it, the full story from Creation and Fall to Redemption and Restoration in the total eclipse of the Sun.

We saw evidence of an intelligent designer in the precise mechanics of the eclipse, and we contemplated God's pleasure in all creation. In this chapter, we discover the story of the Fall in First Contact. Let me explain.

SCIENCE EXPERIMENT #3

See a Crescent Reflected

Materials

- Materials
- A flashlight
- Scissors
- Dark paper, such as a piece of black construction paper
- A writing utensil

Procedure

1. Place the dark piece of paper on a table.
2. Stand the flashlight on its face (its lamp) on top of the paper. (The handle will be facing the ceiling.)
3. Using the writing utensil, trace a circle around the face of the flashlight.
4. Cut the traced circle out of the paper to create a paper disc.
5. Tape the disc over a portion of the face of the flashlight to form a crescent with the lamp.
6. Turn on the flashlight and turn off the lights.
7. Aim the flashlight so it projects light onto the floor.
8. Observe the crescent projected onto the floor. Notice the crescent on the floor opens in the opposite direction of the crescent created by the paper on the face of the flashlight.

Conclusion

In this experiment, we created a crescent with a flashlight and a paper disc. We observed how the crescent we created on the face of the flashlight projected an inverted crescent on the ground.

God's desire for friendship, intimacy, and oneness with us in creation has not changed as described in Genesis, the famous ancient story of beginnings. Jesus expressed this desire for togetherness when he spoke to his friends over 2,000 years ago. On the night of his arrest, sharing the Passover meal with his closest followers, Jesus pleaded, "Remain in me as I also remain in you" (John 15:4a). God desires intimacy with us so much that it grieves him to be separated from us.

God's Grief

During a difficult time in my life, I felt separated from God. I went on a drive in the Texas Hill Country to get away from the world and seek God's face without distraction. As I drove, I cried out to God, "I feel so separated from you."

An image of Jesus on the cross flashed before my mind. I could hear his cry, "My God! My God! Why have you forsaken me?" (Matthew 27:46).

"Oh, Jesus, how you have felt and known the separation I feel to such a greater extreme than I do now!" I answered him.

A silent voice surprised me: *Amy, that was not the first time a separation broke my heart.*

Tears fell down my cheeks as the image of the cross was replaced by God's face when Adam and Eve chose to leave him.

Adam and Eve's Choice to Separate from God

God had given Adam and Eve total freedom in the garden. He said, "You are free to eat from any tree in the garden" (Genesis 2:16, emphasis added). God continued, "But you must not eat from the tree of the knowledge of good and evil, for when you eat from it, you will surely die" (Genesis 2:17).

God named the Forbidden Tree *the tree of the knowledge of good and evil.*

The word knowledge in the original text comes from the

Hebrew root word, *yâda'*.[1] *Yâda'* is the same word used in Genesis 4:1a, "Now Adam *knew* Eve his wife, and she conceived." To have knowledge or to know comes with this connotation of experience and intimacy. For Adam to *know* Eve, we understand that he experienced her and had intimacy with her. For a human to *know* good or evil, we can also understand that a person has had experience with and intimacy with good or evil.

Adam and Eve had known only good, prosperity, excellence, kindness, joy, and pleasure.

Adam and Eve had known only good, prosperity, excellence, kindness, joy, and pleasure. Evil, what separates us from God and is the opposite of good, is an experience they had not known. I think God was saying, *You are free to leave me, and this is the door through which you can walk to leave. To eat from this tree is to know that which is separate from me, to step outside of my pleasure, my joy, my delight, my benefits, my favor, and everything right. But when you separate from me, you will find no good thing. No good thing, no life, exists apart from me. Apart from me is nothing, only death and decay. Do not venture into that way.* I hear God's advisement to avoid this tree in Jesus's words:

> Remain in me as I also remain in you. No branch can bear fruit by itself; it must remain in the vine. Neither can you bear fruit unless you remain in me. I am the vine; you are the branches. If you remain in me and I in you, you will bear much fruit; apart from me, you can do nothing. (John 15:4-5)

But the tempter lied and accused God of lying in Genesis 3:4-5. "'You will not certainly die,' the serpent said to the woman. 'For

1 "H3045 - Yâda' - Strong's Hebrew Lexicon (NIV)." *Blue Letter Bible,* Mar. 1996, www.blueletterbible.org/lexicon/h3045/niv/wlc/0-1/. Accessed 25 Nov. 2023.

God knows that when you eat from it, your eyes will be opened, and you will be like God.'"

But weren't they already like God? Made in his image? Satan deceived Eve as if to say, *That's not true. God is lying; he withholds from you. Life exists apart from God and his way of doing things. Apart from God, you can find pleasure, and you can have life on your own. God withholds something desirable from you, a life independent from him.*

Eve saw the fruit was pleasing to the eye, so she took some and ate it. We depend so heavily on our eyesight. Perhaps we depend on it more than our other four senses put together. Sight has a major weakness: it only allows you to see what's on the surface of objects. On the surface, this tree was "pleasing to the eye" (Genesis 2:9a).

Sight has a major weakness: it only allows you to see what's on the surface of objects.

Adam and Eve chose to sever their ties to God, to separate from him in pursuit of the pleasure they saw with their eyes, in pursuit of their independence, in pursuit of living without God. With the first bite, their minds came to know evil: ideas and thoughts outside of God's good ways. Thoughts like those bring fear and shame, so they hid.

They didn't have the courage to tell God their choice to separate from him, but God knew what happened. Adam and Eve heard his footsteps in the garden. In Genesis 3:9, I hear God express his anguish and desire for them when he called out, "Where are you?" I hear more in the question from the God who knows everything. I hear him asking, Have you decided to separate from me? *Have you decided to leave me? Have you forsaken me?*

"The woman you put here gave it to me," Adam declared (Genesis 3:12).

"The serpent deceived me," Eve answered (Genesis 3:13).

From that point on, Adam, Eve, and all humankind would know evil. This transition from knowing only good to knowing good and evil is the Fall. From that point on, evil spread.

The Spread of Evil

In Genesis 4, we read of Adam and Eve's sons, Cain and Abel. They both gave gifts of sacrifice to God. Abel, a shepherd, brought fat portions from the firstborn of his flock. Cain, a farmer, brought the fruits of the soil. God looked with favor on Abel and his offering but not on Cain. Why did God look with disfavor upon Cain's offering? It had nothing to do with animals or crops. No, God desires mercy, not sacrifice (Hosea 6:6a). God was displeased with Cain's offering because of Cain's heart posture. And we see the fruit of Cain's bitter heart when Cain becomes so angry that he murders his brother.

James 3:16 says, "For where you have envy and selfish ambition, there you find disorder and every evil practice." Envy and selfish ambition may seem like small sins, but they can stealthily grow, leading to confusion, restlessness, instability, and wickedness of all sorts until goodness appears to be nearly snuffed out. Evil has a greedy desire to take over. God warned Cain in Genesis 4:7, "Sin is crouching at your door; it desires to have you."

Intimacy with evil can begin small, like the first tiny bite the Moon takes out of the Sun in First Contact. But from there, it devours, overpowers, and rises to the point of lifelong isolation, bitterness, anger, selfishness, impatience, frustration, and other negative emotions and choices that steal our joy, peace, and relationships. Why would God allow this?

Why Did God Allow Evil to Enter the World?

I often wondered why God made a way for the Fall. Why did he give a choice to Adam and Eve if the consequences would lead to such desolation?

I heard it preached, "God didn't want robots or puppets."

Though this is true, the answer never resonated with my soul.

As I came to know God as a God of freedom, the thought came to my mind, "God didn't want slaves." Though true, neither did this thought completely satisfy my inquiry.

It wasn't until I sat in a break-out session at a women's retreat in 2021 that God gave me an answer that wowed me with his beauty. The leader of that session, Courtney, pointed us to the book *The Four Loves* written by C.S. Lewis. Lewis explores four kinds of love: affection, friendship, eros, and charity.

Affection is best represented by the kind of love a child has for his mother. He needs her to survive.

Friendship is a form of love birthed from a shared interest.

Eros embodies the idea of being in love—a preoccupation with, an adoration of, and a need for another person. Although sexuality comes under the eros kind of love, sex is not its ultimate definition. Remember the Hebrew word *yâda'* used in Genesis 4:1. "Adam *knew* his wife Eve, and she conceived and bore Cain." In this verse, the sexual relation comes with the connotation of intimacy, of knowing and being known. But when Lot's daughters tricked their father into having incestuous sex with them by getting him intoxicated, a different word was used. "Come, let us make our father drink wine, and we will *lie* with him" (Genesis 19:32, ESV). The Hebrew word used here is not *yâda'* but *shâkab.*[2] It refers to the act of sex, an exchange of fluids, with no connotation of intimacy, being known, knowing, or love.

Charity is the fourth love Lewis explores. When you love with charity-love, you choose to put the beloved above yourself because you want to. You give fully even when the one who receives your love has nothing to offer in return. We see charity-love demonstrated in creation. God had no need for the Heavens, Earth, or us. God did not need to create us so he could show or

2 "H7901 - shâkab - Strong's Hebrew Lexicon (NIV)." Blue Letter Bible, Mar. 1996, www.blueletterbible.org/lexicon/h7901/niv/wlc/0-1/. Accessed 25 Nov. 2023.

receive love because God is a triune God: Father, Son, and Holy Spirit. Love exists between the three of them. Love existed before our creation.

So why did God make a way for the Fall? Because he did not withhold anything good from us. He gave us the freedom to love fully. He made us in his image. Like him, we're capable of giving charity-love.

We see this capability exercised in the sacrifice of many throughout history and also in our everyday lives. But the freedom to love fully only comes with the freedom to choose, and with the freedom to choose comes the freedom to hate greatly.

The freedom to love fully only comes with the freedom to choose.

True, God did not want robots.

True, God did not want slaves.

And the truth that pricks my heart is that God gave us a choice because he did not want to withhold any good thing from you or me. God wanted us to have the ability to love with a charity-love, the freedom to choose to love fully like he does.

If you're anything like me, you don't always choose to love. Further, you and I can be blind to the root of our choices. Sometimes I'm like a woman on the path of Totality during First Contact who doesn't pay attention to the shadows on the ground or look up to the heavens with special glasses. I can easily go through the day ignorantly without giving much thought to the state of my heart.

Unaware of the State of My Heart

Sometimes people separate from God because they are intentionally rebellious toward God. Others separate because they are unaware, like wandering sheep—that was me. Even though I struggled with promiscuity, arrogance, selfishness, and alcohol in high school, I considered myself a good person. As a child, my parents told me Jesus died on the cross for my sins, and if I believed in him, I would go to heaven. That didn't make a lot of sense to me, but I wanted to go to heaven, so I believed in him. Not realizing I had enemies, I didn't know evil attempted to take over my life. I thought teenage life naturally included promiscuity, vanity, alcohol, paralyzing anxiety, and suicidal thoughts, just like childhood naturally included bicycles, coloring books, and cartoons.

It wasn't until I was 28 years old, married, and pregnant with my second child that I learned to ask myself a powerful question:

Why?

Why did I say the things I said?

Why did I do the things I did?

As I began to ask myself those questions, it was like looking up at the Total Solar Eclipse with safety spectacles during First Contact. I recognized for the first time that some bad characteristics eclipsed the light that was meant to be inside me.

Asking myself, "Why?" changed everything.

Why did I just say that? *Oh, it's because I'm jealous.*

Why did I just do that? *Oh, it's because I'm selfish and impatient.*

"Why did I just do that?" *Oh, wow, because I want to look better than her in front of this group.*

I discovered I was selfish, impatient, jealous, competitive, hateful, proud, and fearful. I was a sinful woman who lived for the praise of people. I was judgmental, and there was no excuse for it; it was just the core of who I was. For the first time, I began to really understand I wasn't a "good person," even though I never did anything "wrong" according to our culture's standards. I was no longer ignorant of the state of my heart. Asking myself, "Why?" changed everything.

It's Biblical to Ask Why

I never realized how biblical it is to ask yourself a question like "Why?" until I was 43 years old. Preparing to share my testimony in a group setting, I sat down to think it through once again. I realized "Why?" was a question God asked Adam's son in Genesis 4.

Before Cain murdered his brother, Abel, God asked him, "Why are you angry? Why is your face downcast?" (Genesis 4:6).

God knew the answer already, but he wanted to help Cain and lead Cain to examine himself by asking, *Why?* God knew about the wrong, dark thing that wanted to take over Cain's heart, like an astronomer is aware of an upcoming eclipse. God's question, if Cain had answered honestly, would have been like a pair of protective paper glasses to help Cain see the growing black spot.

To this day, asking myself, "Why?" reveals root emotions that drive me.

To this day, asking myself, "Why?" reveals root emotions that drive me.

I'll never forget the most heartbreaking, most revealing answer of all to the question of why I did what I did. The Lord prompted me when I was pregnant with my third child.

As I prayed for Jon to be saved, the Lord asked me, *Why do you pray so fervently for your husband to be saved? Is it not because you want to be married to a 'godly' man? Could you not care less if he went to heaven or hell?*

Oh, it was true! The answer brought me to my knees. I was so very wretched, my heart so very black!

So...What's the Point?

The point is you and I have an enemy who attempts to deceive us and separate us from God and God's good plan for our lives. We don't have to be ignorant of his sneak attack or deceived by his lies.

Like the interesting shadows we see on the ground under the trees during a Total Solar Eclipse, we can pay attention to strange happenings in the soil of our hearts. If we answer yes to any of the following questions, it may be a clue that something is wrong.

1. Am I hiding?
2. Am I paralyzed by anxiety?
3. Do I have dark thoughts?
4. Am I ashamed?
5. Do I feel abandoned?
6. Do I want to isolate myself?
7. Do I think I can do it all on my own?
8. Do I feel confused?
9. Am I aggravated, angry, and impatient?
10. Am I competing with people on my team?

Once we recognize something is wrong, we have the power to ask, *why. Why am I hiding? Why am I competing with people on my team?*

As we can put on eclipse glasses to look up and see the First Contact phase of the eclipse, we can also put on our spiritual "safety spectacles" to discern the state of our heart by asking ourselves the powerful question: "Why?"

When we uncover sneak attacks on our hearts, we don't have to worry. We can keep our heads up, with our eyes turned toward God. As we'll see in the coming chapters, God has a way of intercepting the dark and surprising us with his beautiful light, much like the Total Solar Eclipse in its next phase, Second Contact.

CHAPTER NINE

SECOND CONTACT

The Old Covenant

Ethan loves my daughter Joy. When they started dating, he planned nice dinners at fancy restaurants, went hiking and kayaking with her in the beautiful Texas Hill Country, and took her on lovely, lit horse-drawn carriage rides in downtown San Antonio at night. He planned his proposal with the same heart of pursuit, looking for ways to show his love.

He invited her best friend, Rebekah, to assist with the arrangements. He envisioned large lit letters M A R R Y M E on a hilltop in Wimberly, Texas. He took Rebekah to the crest, where they chose the perfect spot and time of day to capture a gorgeous Texas sunset for engagement photos.

Ethan hired a photographer.

Rebekah called a letter company.

"Yes, ma'am, we set up all the letters and tear them down," the letter company explained.

"We do need them on top of a hill, and there's a stairway to the top. Would you set them up there?" Rebekah asked.

"Yes, no problem. We do all the setup and tear down."

"OK, even if it's on top of a hill?"

"Yes, no problem."

On the evening of the engagement, Rebekah and her friend Ben met the set-up crew, two middle-aged, overweight men, who drove up with the large white letters at the bottom of the stairs in a tiny parking lot.

"Oh, we can't carry the letters up those stairs," they muttered when they stepped out of their truck.

"Seriously?" Rebekah smiled.

"Seriously. We can't do it."

Rebekah and Ben immediately grabbed the M and the A and hauled them up the stairs to start setting up for the proposal. They quickly realized they wouldn't have time to carry the other five letters and properly position them before Ethan and Joy arrived. A family of four on a hike saw them rushing and asked, "Hey, is a proposal about to happen up here?"

"Well, we hope so! If we can get the letters in place in time!"

"Need help?"

"That would be amazing!" Rebekah beamed.

After hurrying down the hill, each adult grabbed a letter and lugged it up the stairs while the two children shared the burden of the fifth letter. With the family's help, they lugged the letters up just in time.

Rebekah and Ben hid when Ethan and Joy pulled up. After leading her blindfolded to the top of the hill, Ethan got down on one knee to propose with a solitaire diamond ring against the backdrop of a Texas Hill Country sunset.

Joy said yes.

God pursues us more extravagantly than Ethan pursues Joy, and I see a symbol of it in the Diamond Ring Effect of the Total Solar Eclipse.

Second Contact - The First Diamond Ring Effect

Just when you think the Sun will be totally snuffed out, plunging you into darkness, Second Contact surprises you, bursting forth in the dreadful dark like a knight in shining armor or like a long-awaited calvary in what you thought was a losing battle.

Seconds before Totality, only moments before the Moon perfectly centers itself in front of the Sun, the eclipse reveals a stunning diamond ring in the sky—a perfect, thin, white circle with a single bright burst of white light shining forth like a magnificent sparkling diamond. No article I read in preparation for the 2017 eclipse mentioned this stunning effect. I couldn't believe my eyes. No words can express its dazzling and unexpected beauty.

The eclipse reveals a stunning diamond ring in the sky—a perfect, thin, white circle with a single bright burst of white light shining forth like a magnificent sparkling diamond.

I learned later that scientists call this stage the Diamond Ring Effect. The term was first coined in 1925 when a Total Solar Eclipse crossed the East Coast of America, and observers claimed they saw an "engagement ring" in the sky.[1]

Those who did not see the ring suggested spectators "must have been tricked by their eyes." Fortunately, E.J. Stein captured a photo of the bedazzling ring, which was published in The New York Times four days after the eclipse, proving its stunning existence.[2]

1 Golub, Leon, and Jay M. Pasachoff. "In the Shadow of the Moon." *American Scientist,* 12 Oct. 2017, www.americanscientist.org/article/in-the-shadow-of-the-moon. Accessed 27 Nov. 2023.

2 "That 'Diamond Ring' in the Sun's Eclipse." *The New York Times,* 28 Jan. 1925, p. 18.

Second Contact exceeds all expectations, and it's at this point that humans begin to cry out, "Wow!" "Diamond Ring!" or "Oh my God!"

What Causes the Diamond Ring Effect?

The craters of the Moon cause the Diamond Ring Effect. Some of those potholes, hollows, cavities, and basins are caused by outside impacts, asteroids, and rocks that have repeatedly bombarded the Moon's surface. Others are caused by volcanoes once active in the Moon's ancient past.[3]

As the Moon's damaged face perfectly positions itself over the Sun, its deepest craters and valleys allow sunlight to pour through them, creating what looks like small beads of light. Immediately before Totality, light pours through the last deep valley for a fleeting moment to create one bead, which shines out like a solitary diamond.[4] If the Moon's surface were perfectly smooth, light would shine equally around it, and no diamond ring would be seen.

The dark side of the Moon, which never faces Earth, is more potted than the face of the Moon we see in the sky.[5] Though the astronauts of the Apollo 8 mission did not set foot on the Moon's surface, they were the first to see the far side of it with their own eyes.[6] The Moon's damaged surface surprised Astronaut Frank Borman when he saw it,

> I don't think anything I'd studied prepared me for the really troubled nature of the lunar surface—it was messed up beyond belief...It was terribly distressed with holes,

3 Burke, James D. "Moon." *Encyclopædia Britannica,* Encyclopædia Britannica, inc., 25 Aug. 2023, www.britannica.com/place/Moon/Distinctive-features. Accessed 25 Dec. 2023.

4 Littmann, Mark, and Fred Espenak. "The Approach of Totality." *Totality: The Great American Eclipses of 2017 and 2024,* Oxford University Press, Oxford, 2017, p. 136.

5 "File:Moon Farside LRO.Jpg - Wikimedia Commons." *Wikimedia Commons,* 6 Mar. 2019, commons.wikimedia.org/wiki/File%3AMoon_Farside_LRO.jpg. Accessed 25 Dec. 2023.

6 Lusko, Levi. "Everyone Is a Moon." *Last Supper on the Moon: NASA's 1969 Lunar Voyage, Jesus Christ's Bloody Death, and the Fantastic Quest to Conquer Inner Space,* W Publishing Group, an Imprint of Thomas Nelson, Nashville, TN, 2023, p. 2.

> crators [sic], volcanic residue, so it was a very interesting first view of a different world.[7]

The gray battered surface of the Moon may appear "messed up beyond belief" on its own, but line it up perfectly with the Sun, and it will astound you, bring you to your knees, and cause you to cry out, "Wow! Oh my God!"

We people can be like the Moon's cratered surface. Each of us is wounded by our own inner turmoil (like volcanoes on the Moon) and also intentionally and unintentionally wounded by other people and circumstances (like asteroids striking the Moon). But align our wounded, messed-up selves perfectly with God so his light might pour through us, and you have a testimony that leaves observers awestruck and encouraged.

Second Contact and the Story of the World

If we connect the story of the world to the phases of an eclipse, Second Contact showcases God's response to the Fall.

As we saw in the last chapter, God's love compelled him to pursue Adam and Eve. His heart went out to his children and his feet followed, stepping into the mess they created as a loving and involved father would. God performed the following actions:

1. God asked them questions for which he already knew the answers leading his children to examine their hearts (Genesis 3:9, 11, 13).
2. God gave a promise, a foretelling of the future: The woman would birth a victor who would ultimately crush the head of this deceiver (Genesis 3:15).

7 Lusko, Levi. "Everyone Is a Moon." *Last Supper on the Moon: NASA's 1969 Lunar Voyage, Jesus Christ's Bloody Death, and the Fantastic Quest to Conquer Inner Space,* W Publishing Group, an Imprint of Thomas Nelson, Nashville, TN, 2023, p. 3.

3. God explained the coming consequences in detail. Remember, Adam and Eve had known only good—pleasure, prosperity, bounty, joy, wholeness, and rightness. Now they had chosen also to know evil, something previously foreign to them—pain, distress, misery, injury, lack, adversity, malignancy, depression, brokenness, and wicked thoughts. This meant that not only would Earth produce thorns and thistles, but Adam and Eve would decay into dust (Genesis 3:18-19).
4. God made the first sacrifice to provide for their needs, sturdy garments of skin, to clothe and protect Adam and Eve (Genesis 3:21). Can you believe it?

God's Response through Sacrifice for Provision

"The Lord God made garments of skin for Adam and his wife and clothed them" (Genesis 3:21). The description of God's provision is given in only one sentence; however, it takes days from the moment you kill an animal to the moment its hide is prepared for wear. The task involves killing, skinning, stretching, drying, scraping, soaking, softening, tanning, and sewing. Making leather is an ancient art at least 7,000 years old.[8] Perhaps the craft originated with this first sacrifice, God's demonstration in the garden. The Moon cannot black out the Sun but rather shows off the Sun's beauty in a stunning diamond ring, so also Adam and Eve's choice to separate from God could not eclipse God's love but rather put the brilliance of his love on display.

The Moon cannot black out the Sun but rather shows off the Sun's beauty in a stunning diamond ring.

God gifted the finished garments to Adam and Eve. Can

8 Britannica, The Editors of Encyclopaedia. "leather". Encyclopedia Britannica, 15 Sep. 2023, www.britannica.com/topic/leather. Accessed 20 October 2023.

you imagine the scene? I envision God holding them horizontally, folded neatly in the palms of his hands outstretched to Eve. I wonder if Eve, now knowing evil but still knowing good, whispered, "Thank you." I wonder if he responded, ensuring her eyes locked on his as he said, "It's my pleasure." I wonder if God and Eve cried when they said goodbye. The fruit was pleasing to the eye, but could it compare to the beauty, kindness, and pleasure of God? At what point did Adam and Eve learn there is no good thing and nothing worth wanting apart from God? At what point did they realize God loved them much more than they ever imagined?

At what point did Adam and Eve learn there is no good thing and nothing worth wanting apart from God?

God Knew and Loved Anyway

God knew from the beginning that Adam would choose to distrust him, to separate from him, yet God decided to create and love him anyway. Let's re-read the warning God gave Adam, placing emphasis on the word when. God warned, "You are free to eat from any tree in the garden, but you must not eat from the tree of the knowledge of good and evil, for *when* you eat of it, you will surely die" (Genesis 2:17, emphasis added). God did not say "*if* you eat of it," but rather he said, "*when* you eat of it."

I think of Adam in his old age, looking back on his life and remembering God's words. I wonder if, like me, the word *when* stood out in Adam's mind, and he realized God knew all along what he would choose, yet God created and loved him anyway. It's God's way of loving: "But God demonstrates his own love for us in this: While we were still sinners, Christ died for us" (Romans 5:8). God's way of loving is displayed in his faithfulness.

God's Faithful Pursuit and Provision

Second Contact is a promise ring in the sky, reminding me of God's vow: "And I will put enmity between you [the deceiver] and the woman, and between your offspring and hers; he [the woman's offspring] will crush your head, and you will strike his heel" (Genesis 3:15). God gave this significant promise in just one sentence; however, it takes the entire Bible to understand how God fulfills his pledge through the offspring of Eve.

God's provision and pursuit continued down Adam and Eve's lineage. Like the faithfulness of the sunrise and sunset, like the consistency of constellations moving across the heavens, God kept watch over and made provision for their descendants with their genealogy carefully recorded in *begat* after *begat.* Ten generations of fathers and their sons are recorded in Genesis 5, starting with Adam and ending with Noah.

> Like the consistency of constellations moving across the heavens, God kept watch over and made provision for their descendants with their geneology carefully recorded in begat after begat.

Noah

When we get to Noah, we find Earth totally corrupted by violence: "The Lord saw how great the wickedness of the human race had become on the earth, and that every inclination of the thoughts of the human heart was only evil all the time. The Lord regretted that he had made human beings on the earth, and his heart was deeply troubled" (Genesis 6:5-6). It broke my heart to think God may have *regretted* making me, so I looked more deeply into the word used in Genesis 6:6. The original text uses the Hebrew word *nâcham,* which means to sigh, to breathe strongly, to draw breath forcibly, to groan, to grieve.[9] God grieved the evil, and don't we admire this about him? Are not our hearts also deeply troubled by evil intent, corruption, and violence on Earth?

Watching the news today can quickly cause chest pain, heavy sighs, groans, and even tears. Don't we want a God who grieves alongside us? And don't we want him to use his power to bring justice and wash away wickedness? He does. He did.

God wiped out the evil generation with a flood and kept his promise to Adam and Eve. God found Noah righteous in his era and proposed a partnership: Build this ark, move inside it with your family, and I will save you when I bring a great flood to wash away a wicked generation on Earth. Noah said *yes* to God's proposal by building the ark just as God commanded him. I have no doubt Noah was amazed by the way the ark kept him safe when it must have seemed that torrents of rain would wash him away. It makes me think of how we are surprised by the Diamond Ring Effect just when we feel some dark disk will completely snuff out the light of the Sun.

The lineage of Adam and Eve continued, with ten more generations of fathers and sons carefully recorded in Genesis 11:10-26 from Noah's son to Abram.

Abram

When we get to Abram, we find God once again working to keep his original promise. "He [God] took him [Abram] outside and said, 'Look up at the sky and count the stars—if indeed you can count them.' Then he said to him, 'So shall your offspring be'" (Genesis 15:5). Abram believed the Lord even though his wife was barren. But when God explained that his offspring would take possession of land large enough to contain a nation, Abram asked

9 "H5162 - Nâcham - Strong's Hebrew Lexicon (NIV)." *Blue Letter Bible,* Mar. 1996, www.blueletterbible.org/lexicon/h5162/niv/wlc/0-1/. Accessed 27 Nov. 2023.

God, "How can I know?" God answered by instructing Abram to prepare to make a covenant according to the culture of the time (Genesis 15:9).

Two thousand years before Christ when a man made a promise, he walked between slaughtered and divided animals as if to say, *If I don't keep my oath, may I lose my own life and become like these torn animals.*[10,11] Abram slaughtered and divided the animals in preparation for the covenant. At sunset, "a dreadful darkness came over him" (Genesis 15:10-12).

A dreadful darkness came over him; these words describe well a Total Solar Eclipse directly before the first Diamond Ring Effect. Remember, before the first Diamond Ring Effect, we experience First Contact, a phase of the eclipse that lasts over an hour. At the end of First Contact, just when you think the Moon will totally snuff out the Sun, you are surprised by the First Diamond Ring Effect and then the pearly crown of Totality. Rather than being plunged into darkness, you are instead stunned with a glorious light show. I wonder if the surprise of the first Diamond Ring Effect is anything like the surprise Abram must have felt when God, in the form of "a smoking firepot with a blazing torch appeared and passed between the

> Rather than being plunged into darkness, you are instead stunned with a glorious light show.

10 Jared T. Parker, "Cutting Covenants," in The Gospel of Jesus Christ in the Old Testament, The 38th Annual BYU Sidney B. Sperry Symposium (Provo, UT: Religious Studies Center, Brigham Young University, 2009). pp. 111-113.

11 An example from an extrabiblical source can be found in an ancient Assyrian treaty, "This head is not the head of a lamb, it is the head of Mati'ilu, it is the head of his sons, his officials, and the people of his land. If Mati'ilu sins against this treaty, so may, just as the head of this spring lamb is torn off, and its knuckle placed in its mouth, [...], the head of Mati'ilu be torn off, and his sons." Pritchard, James B., and Erica Reiner. "Akkadian Treaties from Syria and Assyria/Treaty Between Ashurinirari V of Assyria and Mati'ilu of Arpad." *Ancient Near Eastern Texts. Relating to the Old Testament,* Third Edition with Supplement ed., Princeton University Press, Princeton, New Jersey, 1969, pp. 532–533.

[animal] pieces" (Genesis 15:17). It meant God would put keeping the oath above his own life.

When Abram (also known as Abraham) was 100 years old, and his wife Sarah was 90, God fulfilled his promise and gave them their first and only son Isaac (Genesis 21:3). Through Isaac, God promised to build a nation through which all nations of the Earth would be blessed (Genesis 22:18).

Isaac begat Jacob (Genesis 25:26). God changed Jacob's name to Israel (Genesis 32:28, 35:10-11) and Jacob fathered twelve sons, who formed the twelve tribes of the nation of Israel (Genesis 49:1-28).

The Nation of Israel

The twelve sons of Israel lived in Egypt and grew in numbers (Exodus 1:5-7). Eventually, an Egyptian king, threatened by the multitude of Israelites residing in his country, forced them into slavery and demanded the murder of their baby boys. One mother bravely hid her son in a basket in the reeds of the Nile where Pharoah's daughter bathed (Exodus 2:3-5).

Imagine this mother's surprise when she discovered Pharoah's daughter saved her baby from death and took him in as her own (Exodus 2:10). I imagine it was something like seeing the Diamond Ring Effect of a Total Solar Eclipse in Second Contact. To make a long story short, God kept his promise to Adam, Noah, Abram, and Israel by investing in Moses.

God raised Moses up to lead the Israelites out of captivity. Once freed, God began to establish a nation. Through Moses, God gave the law for this nation, which can be categorized into two forms:

1. The Ten Commandments inscribed in stone, such as "You shall have no other gods before me" (Exodus 20:3) and "You shall not commit adultery" (Exodus 20:14).
2. The rules and regulations written on paper, such as "But if there is serious injury, you are to take life for life, eye for eye,

tooth for tooth, hand for hand, foot for foot, burn for burn, wound for wound, bruise for bruise" (Exodus 21:24-25).

Jesus spoke about the law in both of its forms. Of the Ten Commandments, Jesus said, "You have heard that it was it was said, 'You shall not commit adultery.' But I tell you that anyone who looks at a woman lustfully has already committed adultery with her in his heart" (Matthew 5:27-28). What does this mean? It means God is interested in us at a much deeper level than our outward behavior. He's interested in our hearts. He is also teaching us what it means to live up to his standard, to be like him. He never has a hateful or wrong thought. Who can live up to his standard in their own power? No one.

Jesus also spoke about the rules and regulations on paper. I often wondered about this. At first, it seemed to me that Jesus contradicted the law when he stated, "You have heard that it was said, 'Eye for eye, and tooth for tooth.'...But I tell you, love your enemies and pray for those who persecute you" (Matthew 5:38, 44). Did God change?

When I asked God about this seeming contradiction one day, it dawned on me: God was establishing a nation on Earth with a justice system. The rules and regulations were instructions for judges, to teach them how to execute justice fairly in this new nation. Jesus, however, did not come to establish a nation on Earth but rather to invite us into God's nation. In God's kingdom, we are not called to judge but to love. God is the judge of his kingdom, and we are not qualified to do his job.

As citizens of a country, we long for justice to prevail. God's plan for the nation of Israel was to serve as their judicial, legislative, and executive branch. Isaiah said, "For the Lord is our judge, the Lord is our lawgiver, the Lord is our king; it is he who will save us" (Isaiah 33:22). I want to be a part of a country that is good, with fair laws, honest judges, and brave, right-minded leaders. I don't want to live in a nation that allows a thief to take out my daughter's eye and get away with it. Our bones cry out for justice. Yet the judges of Israel failed to live up to the standards of rightness given by the law that came through Moses. Time and again, the nation turned away from God—their written history riddled with the words, they "did evil in the eyes of the Lord."[12]

People Can't Seem to Faithfully Wear God's Ring

God kept his promise to Adam and Eve by birthing a nation through Noah, Abraham, and then Moses. God made a proposal to the nation, "Walk in obedience to all that the Lord your God has commanded you, so that you may live and prosper and prolong your days in the land that you will possess" (Deuteronomy 5:33). But people could not seem to take God up on his offer. No one could perfectly follow his good ways.

God made a proposal to the nation.

Everyone is like this. We are all sheep who have wandered away and chosen our own path. No one follows the commands of God perfectly from the heart. The Law of Moses established a nation with a fair justice system, and it also put on display our inability to live up to his perfect standards.

My Need for God's Purification

As I shared in the previous chapter, I learned to ask myself, "Why?" as a young mother. This question helped me discover the root motives behind my behaviors and led me to a deep understanding of my great need for a savior. I discovered I was compelled by selfishness, jealousy, pride, anger, and bitterness. Furthermore, I didn't have the power to change; it seemed to be the core of who I was.

12 2 Kings 21:16, 2 Chronicles 29:6, Judges 3:7, Judges 3:12, Judges 10:6, 1 Kings 14:22, 2 Chronicles 22:4

Shortly after this discovery, we moved into our red brick one-story house. Even though I was prideful, judgmental, bitter, angry, and selfish, God pursued me. Laurie, the neighbor who lived two doors down and hosted the annual Christmas cookie exchange, invited me to a Bible study in her home. I learned 1 John 1:9: "If we confess our sins, he [God] is faithful and just and will forgive us our sins and purify us from all unrighteousness." Wow. Purify us from all unrighteousness! What a gift! God was offering *more* than forgiveness. He was offering *purification* plus he was offering purification from *all* unrighteousness! Freedom from wrong thinking, speaking, and behaving! I took it for truth and rejoiced in seeing there was a way out of this sinful state I was in.

"Laurie, can you come over to my house today?" I asked her after Bible study. "I only need five minutes of your time."

"Sure, I'll come after everyone leaves."

When she arrived, I didn't even invite her to sit down, and we stood in the kitchen.

"Laurie, I need to confess something to you. It's ugly and embarrassing, and I'm only telling you because I'm standing on 1 John 1:9 that says if we confess our sins, Jesus is faithful and just to not only forgive us but also purify us, and I need that. I'm desperate."

She stood in front of my white refrigerator, waiting for me to continue while holding eye contact.

"In our Bible study," I continued, "I feel jealous of the other women, and I'm actually competing to look smarter and more spiritual than everybody else there. I hate that, and I'm only confessing it in hopes Jesus will give me freedom from it."

I imagined she would respond with an admonishment, like, *Oh, friend, quit feeling jealous. It's not a competition, and those ladies are really nice.*

"Oh, I totally get it!" she declared with fire in her eyes. She hugged me up, overjoyed with my confession.

Her response surprised me, much like the Second Contact of the Total Solar Eclipse surprised me. She loved me and gave me a picture of God's love for me despite my embarrassing confession.

So...What's the Point?

The point is that God loves and pursues you and me despite our inability to follow his good ways in our own power perfectly. We're like the cratered Moon, full of imperfections and sometimes "messed up beyond all belief."[13] But when we let God's light pour through our wounds, his extravagant love astounds us. The first Diamond Ring Effect is like a promise ring in the sky, reminding us of God's response to the pervading darkness that came to Earth through the Fall.

God loves and pursues you and me despite our inability to follow his good ways in our own power perfectly.

God pursued Adam and Eve in the garden when they turned from God and went their own way. He didn't pursue them to punish them but to provide for them as they set out on their own. He also gave them a promise: God would provide a victor through their offspring who would defeat their deceiver. The spreading darkness of evil would not overwhelm the Earth, no matter how many times it tried.

Evil tried to overwhelm Earth in Noah's day. When the entire human race became so wicked that every leaning of every heart was only evil, God pursued one man in the family line.

13 Lusko, Levi. "Everyone Is a Moon." *Last Supper on the Moon: NASA's 1969 Lunar Voyage, Jesus Christ's Bloody Death, and the Fantastic Quest to Conquer Inner Space,* W Publishing Group, an Imprint of Thomas Nelson, Nashville, TN, 2023, p. 3. (Astronaut Frank Borman described the Moon with these words.)

God counted Noah righteous and gave Noah the opportunity to partner with him, be rescued, and be included in God's work to preserve the human race.

God kept the promise alive in the family line when he pursued Abram. When Abram asked God to help him believe and dreadful darkness descended upon him, God blazed through the darkness as a fiery torch, passing through the animals Abram cut in half to show that the promise depends on God, not us.

When Egypt oppressed Abram's growing family and attempted to diminish their numbers, God rescued them. God brought them out of slavery and established the nation of Israel. With good commands, rules, and regulations, God showed his justice and rightness. We are unable to be as just and right as he is in our own strength.

God pursues humanity more extravagantly than a good man pursues a woman he loves and wants to marry. Nothing can get in his way. Even our inability to live up to his standards does not stop his pursuit. That's what we'll see in the next chapter, the climax of God's pursuit.

CHAPTER TEN

TOTALITY

Jesus Christ

One summer, I walked through one of my favorite aisles at our local grocery store, the one with pens and paper, with one of my favorite people, my 19-year-old daughter, Glory. I needed a new spiral notebook, but they only offered plain options in basic colors. I laughed half-heartedly and lamented the selection, "Oh, the little girl inside me so very much wanted one with a pretty design on the front. Oh well, she'll have to settle for this ugly one."

The following morning, I found the spiral sitting on the kitchen table with a hand-painted cover. A white wooly sheep with a smooth brown face and smooth brown ears curled up to rest on the dark black cover of the spiral. She kept her neck up with her head turned toward her tail and her eyes closed. From the top of her body, quarter-inch swirls of blended red and white poured forth and branched out like extra curly corn stalks amid a background of blended blues—light, aqua, and midnight. White stars and spots dotted the blue backdrop like crepe myrtle flower petals scattered on the sidewalk.

What a beautiful surprise Glory had created for me! Totality is like that; it's a beautiful surprise.

Totality Phase of the Eclipse

Totality occurs immediately after the Diamond Ring Effect of Second Contact. During this climax phase of the eclipse, the Moon perfectly blocks the Sun's brilliance, providing a phenomenal view of the Sun's outer atmosphere. White, ethereal, fiery flames spread forth from the Moon's edges. My husband describes it as a blowtorch blasting on the back of the Moon. I describe it as a living crown because of the active movement of the fire that forms the wreath.

White, ethereal, fiery flames spread forth from the Moon's edges.

Totality delivers not only this glowing crown but also the appearance of a simultaneous sunrise and sunset, which means we see beautiful red skies on the horizon not only in the east or the west but all around us. Because I was nestled in the trees during the 2017 eclipse, I missed this effect, but observers who have a 360-degree view of the horizon may have the opportunity to experience it. It seems strange. How can an eclipse cause this effect? It helps to understand what makes the horizon reddish at sunrise and sunset.

The red hues we see on the horizon at dawn and dusk are caused by sunlight taking a longer path through the Earth's atmosphere to get to our eyes. The atmosphere is primarily comprised of nitrogen and oxygen, which are good at reflecting and redirecting blue light.

When the Sun is directly overhead (at noon, for example),

sunlight takes its shortest path through the Earth's atmosphere to reach our eyes. We're immersed in the blue light reflected and redirected by nitrogen and oxygen in our troposphere (the layer of our atmosphere that is closest to Earth's surface, reaching up approximately seven miles).[1] This makes the sky look blue.

When sunlight comes from the horizon, it takes a longer path through the atmosphere. By the time the sunlight reaches our eyes, much of the blue light has been scattered out while the orange and red hues continue the long journey to our eyes, giving the sunset and sunrise their fiery colors.[2]

From the path of Totality, we stand under the Moon's shadow, which is approximately 115 miles wide.[3] Sunlight coming from the edge of the shadow takes a long enough path through Earth's atmosphere to scatter much of the blue light. The crimson colors persevere, reach our eyes, and give us the 360-degree sunset effect.

This phase of the eclipse is one you wish to abide in for a while, and what's more wonderful about it is it's the only time onlookers are allowed and even encouraged to remove their safety glasses.

Safety Glasses Off

Did you ever try to look at the Sun when you were a little kid? I did. I really wanted to see it. I didn't know it was unsafe to look straight at the Sun, so I'd put my hand up above my eyes like the brim of a baseball cap, squint as tight as I could, and attempt to peer directly at it. Every time, the blinding brightness would command my eyes to turn away. I could never get past the Sun's brilliance to get to the Sun. I learned later that looking at the

1 "Atmosphere." *Education,* National Geographic, education.nationalgeographic.org/resource/atmosphere/. Accessed 20 Jan. 2024.

2 Corfidi, Stephen F. "The Colors of Sunset and Twilight." National Oceanic Atmospheric Administration, Sept. 2014, www.spc.noaa.gov/publications/corfidi/sunset/ Last Accessed 19 Oct. 2023.

3 Tavernier, Lyle. "The Science of Solar Eclipses and How to Watch with NASA - Teachable Moments." *NASA,* Jet Propulsion Laboratory, 19 Oct. 2023, www.jpl.nasa.gov/edu/news/2023/9/28/the-science-of-solar-eclipses-and-how-to-watch-with-nasa.

Sun can actually burn your eyeballs; besides, it always gave me a headache, so I quit trying.

We can look directly at the Sun if we wear special glasses, lenses so dark you can't see your hand in front of your face when you wear them. These lenses must meet special requirements defined by a group of experts around the globe known as the International Organization for Standardization (ISO). Only glasses marked with "ISO12312-2" allow us to look at the Sun safely.[4] But there is one time and one place when the Sun allows us to look directly at it without these special glasses: during Totality from the path of Totality.

Directly under the Moon's shadow during the phase of Totality, the Sun's brilliant photosphere (the surface of the Sun) is totally covered. With the brilliance of the surface shrouded, we can freely gaze upon the Sun's chromosphere (a layer of plasma directly above the photosphere) and corona (the Sun's outer atmosphere). There's no other circumstance that allows it. Scientists take advantage of this special situation to observe the Sun and the universe, which has allowed us to discover truths we might otherwise not know.

Totality Reveals Things Unseen

The Moon's attempt to cover the Sun only reveals her glory and the mystery of our universe in fresh new ways. Totality is like an experiment set up by a grand teacher to show us aspects of the universe we would otherwise not be able to see. Totality gave us a first view of solar prominences (red-glowing plasma extending outward from the Sun's surface) and the corona (the sun's atmosphere). The arrangement of Totality also provided a special situation that enabled scientists to prove Albert Einstein's famous theory of general relativity, which says that gravity is caused by massive objects curving space and time.

4 "Safety - NASA Science." NASA, NASA, science.nasa.gov/eclipses/safety/. Accessed 23 Oct. 2023.

Solar Prominences and the Compositional Makeup of the Sun

Solar prominences (plasma extending outward from the Sun's surface)[5] look like flames of fire looping out from the edge of the Sun. I had an opportunity to see them once with my own eyes in the summer of 2022 when a local astronomy club[6] made my childhood dream come true and let me stare at the Sun through their telescopes with protective solar filters. Scientists first discovered them during the Total Solar Eclipse of July 8, 1842, but no one knew exactly what they were. Most guessed they belonged to the Sun, others guessed they belonged to the Moon, and still others thought they were only an optical illusion.[7]

The Moon's attempt to cover the Sun only reveals her glory and the mystery of our universe in fresh new ways.

Furthermore, the pearly incandescent crown of this eclipse piqued the curiosity of scientists, who, at the time, did not know what the corona was. Francis Baily, a brilliant banker who later became a distinguished astronomer, experienced Totality for the first time during this 1842 eclipse. He described it with a solid command of language,

> I was astounded by a tremendous burst of applause from the streets below, and at the *same moment* was electrified at the sight of one of the most brilliant and splendid

5 "What Is a Solar Prominence?" *NASA,* NASA, 6 Mar. 2015, www.nasa.gov/image-article/what-solar-prominence/. Accessed 6 December 2023.

6 Thank you to University of Texas in San Antonio (UTSA) space physics doctoral student Caleb Gimar and the UTSA Astronomy Club for sharing their passion for amateur astronomy with me. And thank you to Rob Thorpe, Program Manager at Southwest Research Institute, for setting up the event.

7 Aitken, Robert G. "A TOTAL ECLIPSE OF THE SUN." *Publications of the Astronomical Society of the Pacific,* vol. 29, no. 167, 1917, pp. 25–40. *JSTOR,* www.jstor.org/stable/40710505. Accessed 7 Dec. 2023

> phenomena that can well be imagined. For, at the instance the dark body of the moon was *suddenly* surrounded with a *corona,* or kind of bright *glory,* similar in shape and relative magnitude to that which painters draw round the heads of saints.[8]

According to Fred Espenak, also known as "Mr. Eclipse," and science-writer/eclipse-chaser Mark Littmann, Baily's ability to colorfully describe his observations of this eclipse is known to have "generated fervor" for solar physics (the study of the Sun), and also known to have "founded the industry of eclipse chasing."[9] The resultant astronomical community of eclipse chasers soon discovered Totality provides a unique lens that allows us to uncover new revelations about our universe.

Don't know what something is? Heat it up, scatter the light from its fire with a prism, and each color you see maps to a specific element of the periodic table.

With the advent of new and improved photographic processes, scientists were able to make fresh discoveries about the Sun in the eclipses that followed. Photographs taken during the eclipses on July 28, 1851, and July 18, 1860, proved the fiery protrusions seen approximately ten years earlier did indeed belong to the Sun.[10,11] Additionally, a new invention, a scientific instrument called a spectroscope, allowed scientists to determine what these extensions were made of.

8 Littmann, Mark, and Fred Espenak. "The First Eclipse Chasers." *Totality: The Great American Eclipses of 2017 and 2024,* Oxford University Press, Oxford, 2017, p. 87.

9 Ibid., pp. 87-88.

10 Ibid., pp. 90-93.

11 Aitken, Robert G. "A TOTAL ECLIPSE OF THE SUN." *Publications of the Astronomical Society of the Pacific,* vol. 29, no. 167, 1917, pp. 25–40. JSTOR, www.jstor.org/stable/40710505. Accessed 7 Dec. 2023.

Spectroscopes can be thought of as prisms, or suncatchers, that scatter white light into its various rainbow of colors: reds, greens, and blues.[12] In 1859, German physicist Gustav Kirchoff discovered and demonstrated that every element gives off a specific color of light when heated.[13] Don't know what something is? Heat it up, scatter the light from its fire with a prism, and each color you see maps to a specific element of the periodic table. Isn't that fascinating? Scientists gathered for the August 18, 1868 eclipse with their new spectroscopes in hand.

They discovered the flaming fingers of the sun (the solar prominences) were primarily comprised of hydrogen. With sparked interest in learning the compositional makeup of the Sun, English astronomer Norman Lockyer discovered a new element yet to be identified in the periodic table and suggested the name *helium* (after the Greek word *helios,* which means Sun).[14]

Not only did Totality pave the way for us to discover a new element and the compositional makeup of the Sun, but it also provided the perfect circumstance to prove Einstein's theory of general relativity.

Confirmation of Einstein's Theory of General Relativity

Albert Einstein published his theory of general relativity in 1915. Four years later, a Total Solar Eclipse provided the conditions necessary to confirm it.

Over 200 years earlier, Sir Isaac Newton had theorized that gravity was a force of attraction between objects, a force dependent on the mass of the objects and the distance between them. Einstein theorized something different. Einstein imagined space and time

12 "Spectroscopy: Reading the Rainbow ." *Hubblesite,* NASA, 30 Sept. 2022, hubblesite.org/contents/articles/spectroscopy-reading-the-rainbow. Accessed 6 December 2023..

13 Britannica, The Editors of Encyclopaedia. "Gustav Kirchhoff". Encyclopedia Britannica, 13 Oct. 2023, www.britannica.com/biography/Gustav-Robert-Kirchhoff. Accessed 21 October 2023.

14 Littmann, Mark, and Fred Espenak. *Totality: The Great American Eclipses of 2017 and 2024,* Oxford University Press, Oxford, 2017, pp. 93-102.

joined together to create a new concept he called spacetime.[15] He theorized that gravity was the result of matter bending spacetime and thus compelling the movement of objects. It's not easy to imagine. When Einstein published his theory, he warned that no more than twelve people on Earth would understand it.[16] Gravity is a difficult idea to think about. As a child, I never wondered why anything fell to the ground when you let it go. Did you? But I was puzzled when I pondered planets and stars "floating" in outer space. Heavy things don't float. *How do they just float out there?* I wondered.

Perhaps the easiest way to grasp Einstein's theory is to consider a trampoline, a marble, and a bowling ball. Roll a marble across a flat trampoline, and the marble will roll straight across to the other side. But set a bowling ball in the middle of the trampoline and notice how it changes the shape of the trampoline—it's no longer flat. This is similar to the way the Sun bends spacetime. Now, roll a marble across the trampoline and notice how it no longer travels straight to the other side but rather curves and rolls toward the bowling ball. Roll the marble just right, and it will circle the bowling ball the way Earth orbits the Sun until friction puts a stop to it.[17]

Newton's theory is not wrong and does explain how planets orbit the Sun, but if Einstein is correct, gravity would affect not only the path of planets but also the path of something without mass—light. Einstein predicted that our Sun would warp spacetime in such a way as to bend the path of starlight.

Because we can't see stars during the day when our side of Earth faces the Sun, the theory seems impossible to test. But, a Total Solar Eclipse provides a way that allows us to see stars in the middle of the day and thus examine the effect of the Sun on the

15 Einstein, Albert. "Albert Einstein on space-time". Encyclopedia Britannica, 6 Sep. 2016, www.britannica.com/topic/Albert-Einstein-on-Space-Time-1987141. Accessed 5 December 2023.

16 "Lights All Askew in the Heavens." *The New York Times,* 10 Nov. 1919, p. 17.

17 Burns, Dan. "Gravity Visualized." YouTube, Physics Teacher SOS , 10 Mar. 2012, www.youtube.com/watch?v=MTY1Kje0yLg. Accessed 5 Dec. 2023.

path of starlight. Would the Sun warp spacetime in such a way as to bend starlight as Einstein predicted? British astronomer Arthur Stanley Eddington planned to answer this question by making careful observations from the path of Totality during the Total Solar Eclipse of May 29, 1919.

Eddington took photos of the Hyades star cluster, the cluster closest to the Sun forming the V-shape in the face of the constellation Taurus the Bull. He knew the exact location of each star in the cluster at night, when starlight traveled in a straight line from their locations to Earth, unaffected by the Sun. He traveled to Príncipe, a tiny island off the West Coast of Central Africa, in hopes of photographing the star cluster during the day when the Sun stood between Earth and the cluster. If the Sun bent spacetime, the stars would appear in different locations in the sky. They did.[18]

The Total Solar Eclipse revealed evidence that Einstein's imagined fabric of spacetime exists.

Einstein's prediction proved true. Six months later, on November 10, 1919, The New York Times published the news in large font, "Lights All Askew in the Heavens…Einstein Theory Triumphs."[19]

The Total Solar Eclipse revealed evidence that Einstein's imagined fabric of spacetime exists. Simultaneously, the eclipse replaced the Sun with a strange black hole in the sky.

The Sun Is Still There

Although some people are drawn to the phenomena of Totality and wish to experience it again and again, not everyone does. The eclipse is strange; I dislike most photos of Totality. Our sky is not

18 Pattison, Darcy, and Peter Willis. *Eclipse: How the 1919 Solar Eclipse Proved Einstein's Theory of General Relativity.* Mims House, 2019.

19 "Lights All Askew in the Heavens." *The New York Times,* 10 Nov. 1919, p. 17.

meant to have a black hole in it. This black hole can bring feelings of grief, fear, and overall distress—perhaps especially to those who feel they have a black hole in their hearts due to the loss of a loved one, depression, infertility, addiction, job loss, sickness, divorce, trauma, or some other source of grief. But what I want us to remember about that black hole in the sky is that the Sun is still there. We see evidence of its presence in its pearly chromosphere radiating from the edges of the black hole.

The clouds and black holes of life can block our vision for a time. We lose sight of God and wonder if he's there or if he's good. He is. He's always there, and he's always good, even when we can't see him. After the Fall and before Jesus came to Earth, we couldn't see God (Exodus 33:20). God spoke through the prophets like Noah, Abram, and Moses (Hebrews 1:1). But when Jesus came, God spoke through him, and we saw the image of the unseen God in him (Hebrews 1:2-3, Colossians 1:15). And that's what I see in the Totality phase of the Total Solar Eclipse, God in the flesh.

Totality and the Story of the World

Totality is the central phase of the Total Solar Eclipse and we can map it to the middle of the story of the world: Jesus on Earth, crucified, buried, and raised to life. The crown in the sky reminds us of the kingship of Christ. In the story of the world, we recognize him as a center point in history through the Gregorian calendar, the most widely used calendar in the world.

In ancient times, historical records primarily referenced regnal years, years relative to the reigning king of the time.[20] For example, historical records might contain the words "in the fourth year of

20 Wiesenberg, E.J. , Woodhead, A. Geoffrey , Ziadeh, Nicola Abdo , Wu, Shih-ch'ang , Filliozat, Jean L.A. , Momo, Hiroyuki , Thompson, Eric , Helck, Wolfgang and Rowton, Michael B.. "chronology". Encyclopedia Britannica, 1 Aug. 2022, www.britannica.com/topic/chronology. Accessed 7 December 2023.

Solomon's reign."[21] The Gregorian calendar divides history into two epochs: before the conjectured year of the conception of Christ and after. The year 512 B.C. references 512 years before the reckoned birth of Christ, where B.C. stands for "before Christ." We could also note this as 512 B.C.E. (Before Common Era). As I write this, we're in the year A.D. 2024, where A.D. is short for the Latin term anno Domini, which means "in the year of our Lord." We could also use the notation 2024 C.E. (Common Era). An astronomical dating system introduces the idea of a year 0 and uses negative and positive numbers to categorize the two epochs. 1 B.C. is year 0, and 512 B.C. translates to year -511. As you may guess, A.D. 2024 is simply 2024.[22]

Jesus remains the defining marker, the center point of humanity's timeline.

Regardless of your chosen notation, Jesus, often referred to as the King of kings, remains the defining marker, the center point of humanity's timeline. Before him, we count the years down. Since him, we count up. What other king has had such power over the calendar of the world? None.

Jesus Reveals God

Like the Moon covers the Sun and reveals its corona, Jesus wraps the brilliance of God in flesh to reveal his compassion, justice, love, and power over death. When we study the walk of Christ on Earth, we get to see and know God's heart for people.

In Luke 7:11-17, Jesus went to a town called Nain with his disciples and a large crowd. As he approached the town gate, pallbearers carried a dead person out—the only son of his widowed mother. A large crowd went with her. When Jesus saw

21 1 Kings 6:1

22 Espenak, Fred. "Year Dating Conventions". NASA, 25 Feb. 2008, eclipse.gsfc.nasa.gov/SEhelp/dates.html. Accessed 7 December 2023.

her, his heart went out to her. You couldn't know it until you saw the body of Christ follow his invisible heart straight to her. The leaders of these two crowds met and stood face to face. I imagine they communicated much through tear-filled eyes. From his heart, his mouth spoke an imperative, "Don't cry." Perhaps the words puzzled the people. One filled with such compassion would not command a grieving mother to stop crying. But he followed the puzzling imperative with an action, a reason not to weep.

He went up and touched the bier they carried the dead son on. The pallbearers paused. In Jewish tradition, a funeral procession proceeded with three to seven pauses, deliberate stops for the express purpose of making room to lament the loss.[23] But this stillness was not meant to make room to know lament. This stillness was meant to make room to know God. With words, Jesus commanded and created, "I say to you, get up!" The dead man sat up and began to talk, and Jesus gave him back to his mother. No one asked him to do that. In the story, we see his compassion, love, and power over death manifest. I doubt the pallbearers or the grieving mother expected this turn of events. It's so surprising. I also think it might have been difficult for people in distant towns, who heard the story, to believe it.

Jesus Surprises Us and Exceeds Our Expectations

Regardless of how many articles you read about Totality, you will not fully comprehend its splendor and glory until you experience it for yourself. No description you read, see, or hear in advance will detract from its surprising effect. So it is with Christ. I can tell you about Jesus all day and night, but until you experience his love for yourself, you cannot understand him or why I love him. Totality is staggering. The death and resurrection of Jesus is more staggering. He's known for exceeding the expectations of those who follow him closely.

---

23 Lamm, Maurice. "A Jewish Burial and Procession - Chabad.Org." *Chabad.Org*, Chabad-Lubavitch Media Center, www.chabad.org/library/article_cdo/aid/281569/jewish/A-Jewish-Burial-and-Procession.htm. Accessed 8 Dec. 2023.

Jesus explicitly told his closest followers several times he would be put to death and then raised to life. In Luke 9:22, he said, "The Son of Man[24] must suffer many things and be rejected by the elders, the chief priests and the teachers of the law, and he must be killed and on the third day be raised to life." Again, in Luke 18:31-33, Jesus told his disciples that when they went to Jerusalem, they would mock him, insult him, spit on him, flog him, and kill him but on the third day he would rise again. Despite his clear description, his disciples didn't comprehend what he meant (Luke 9:45, 18:34).

Even when circumstances began to play out exactly as he said they would, the disciples still didn't understand, as evidenced by their reaction. They forsook Jesus and ran away when the elders and chief priests seized him in the darkness of night (Matthew 26:55-56). Exactly as Jesus described, the religious leaders rejected him (Luke 22:66-71) and delivered him over to the Romans, who flogged, mocked, spit on, and crucified him (Matthew 27:26-44).

As Jesus hung on the cross, only a handful of his best friends and family stood nearby. Four names are mentioned: his disciple John and three women named Mary, his mother, his aunt, and Mary Magdalene (John 19:25). It sounds like the end of First Contact when it appears darkness will overcome the Sun and no one can imagine the brilliance that's coming next.

Why did Jesus have to die anyway?

When you think of the crucifixion of Christ, you may wonder, *Why did Jesus have to die anyway?* I wondered that for many years and didn't understand until I reached my 30s. I'll share that in

24 Erbaugh, Matt. "All the Stories Are True: Jesus Is the Son of God // Matt Erbaugh." *CrossBridge Community Church,* 16 July 2023, crossbridgecommunitychurch.com/sermons/all-the-stories-are-true-jesus-is-the-son-of-god-matt-erbaugh/. Accessed 7 Dec. 2023. (The term *Son of Man* means human being and is first mentioned in Daniel 7:13-14 to describe a human being who is also God. In this sermon, Pastor Matt Erbaugh provides an in-depth explanation.)

the next chapter. For now, let's discover what happened after the crucifixion.

Even though Jesus made it clear he would rise on the third day, the disciples remained in hiding while some of the women went to his tomb with the intention of anointing his corpse with spices and perfumes. Knowing the tomb was guarded and covered with a large stone, they wondered, *Who will roll the stone away for us?* Perhaps they hoped the guards would have the mercy and muscle to move it and let them in. The women certainly did not expect the stone to be rolled away upon arrival. I imagine they hesitantly and reverently entered the open tomb when a young man dressed in white utterly astounded them. "Don't be alarmed," he spoke. "You're looking for Jesus, but he has risen!" Thrown into a state of blended fear and wonderment, they ran out as fast they could (Mark 16:1-8).

The eclipse symbolically depicts Christ's burial place in the sky. The Sun is the grave of Christ, and the Moon its stone. The stone is perfectly positioned over the mouth of the grave only shortly after to be rolled away. I wonder if gleaming white light poured around the edges of the tombstone when Christ came back to life.

The Sun is the grave of Christ, and the Moon its stone.

When the women described their encounter to the disciples, the disciples did not believe them even though their words precisely aligned with the prediction Jesus had given many times (Luke 24:9). I don't blame them. Who can believe a resurrection without experiencing it himself? Besides, it's no doubt the disciples were overwhelmed with grief at the loss of their beloved friend, confusion after the crucifixion of their hope, and perhaps ashamed that they had abandoned him. Like a black hole in the sky at noon, nothing made sense. But Peter ran to the tomb to investigate. Bending over, he saw the linen that previously wrapped the corpse and went away wondering what had happened (Luke 24:12).

Do you think Peter felt a bit like those who wondered if Einstein's idea about the invisible fabric of space and time was true before it was proven? Perhaps. But I imagine when Jesus appeared to his disciples in the flesh post-death, their enthusiasm greatly exceeded the excitement of the scientific community when Arthur Stanley Eddington proved starlight bent. (Luke 24:36-37).

Humanity on Earth seems to bend God's fabric of mercy and grace. The grand and invisible God became small, an object of flesh that gravitated to the object of Earth, compelled by the fabric of his love and goodness. Who can discern it until he experiences it for himself?

Because of his death and resurrection, we died and were born again. He keeps surprising us.

Someone else might die for me. My mom and dad would. Jon would. But would any of them have the power to conquer death and come back to life? An interesting effect of Totality is the appearance of simultaneous sunrise and sunset, which serve in literature as powerful metaphors for birth and death. We see the same interesting effect in Christ on Earth. Because of his death and resurrection, we died and were born again. He keeps surprising us.

My Unexpected New Friends

We watch videos and view photos of Totality, yet none of these measure up to the real event. When we experience Totality for ourselves, it exceeds all expectations. God is like that. The longer I walk with him, the more he exceeds my expectations. For example, he used this eclipse to give me amazing new friends.

In March 2020, my dad called, "Amy, I ate a hot dog with a great man today. He wrote a book, and I want you to read it." He gave me the title, and I immediately ordered the book, *How I Know: A Story to Strengthen Your Faith* by Mario and Danielle Lopez.

One year later, as I talked to God about the book you're reading right now, I heard God say, "I want you to include the man who wrote that book."

"Yes, sir, but I don't remember the name of that man or the name of that book. I'll need help to find him," I responded. I put it on my mental to-do list and imagined I would look in the history of my Amazon shopping cart in the coming days to find the author's name and start a Google search from there.

Three days later, I attended an event that included multiple churches in town. We sat in a gymnasium at large round tables with eight chairs each. During a break, I stood up and scanned the room. The man who wrote that book stood at the other side of the gym! I had never met him before, but I recognized him from the back cover of his book. He was easy to spot, with a missing right arm and a black patch over his right eye. I couldn't believe it. Almost as good as Totality, God had exceeded my expectations by helping me find him fast.

Mario's Unseen Battle

In the book he and his wife Danielle co-authored, *How I Know*, Mario shares his story of serving as a combat engineer in Afghanistan, where he was severely wounded. When an improvised explosive device hit him, he lost his right arm and four fingers on his left hand. After this tragedy, he came up against two near-death experiences, and the second one reminds me of the symbolism depicted in the Total Solar Eclipse.

Recall the progression of the eclipse: Darkness creeps in,

slowly takes over, and just when you think it's going to snuff out your light source at the end of First Contact, Wow! Your light source surprises you as it bursts forth in the form of a diamond ring and then morphs into a living crown, a light that reveals truths about the universe you could not have otherwise known.

Mario underwent outpatient surgery to support the reconstruction of his face. Everything appeared to have worked well, but something went terribly wrong, leading to an extreme loss of blood, a bad reaction to a blood transfusion, a 108° fever, collapsing veins, and convulsions. Mario describes how the room suddenly turned silent when busy doctors, nurses, and his praying mother disappeared. No one remained except him when three entities were revealed: one like a strong man who represented pride, one like a woman who represented lust, and one like a monkey who represented greed. The greedy one hopped onto his chest, stared into his eyes, and knocked on his head to see if Mario was still alive. Cursing, mocking, and cheering his nearness to death, it seemed the threesome had him in their grip until, Wow! An armored angel of light carrying an ancient sword crashed through the ceiling. The ugly entities immediately vanished, and Mario woke up, fully alert.

The failed outpatient surgery was like the First Contact of the Total Solar Eclipse; the three hateful entities eager for his death were like the end of First Contact when you think darkness will completely consume you. The angel of light bursting forth through the ceiling can be likened to the surprising diamond ring. The angel's power and authority over the hateful beings were like the glorious crown of Totality, revealing to Mario his need to get right with the good, strong, powerful God of light.

So...What's the Point?

The point is we get to see and know God through Jesus, who is God in the flesh. If you ever feel confused about who God is and it seems impossible to know or understand him (like a child trying to peer through the brilliance of the Sun), look at the stories of Jesus on Earth. In these stories, we can clearly see his love and compassion for people. The pearly crown of Totality points to the King of kings, maker of heaven and earth.

Totality reveals new discoveries of the Sun—its compositional makeup and its power to warp space and time.

Totality reveals new discoveries of the Sun—its compositional makeup and its power to warp space and time. Jesus on Earth, crucified, buried, and raised to life, gives us new discoveries about God—his compositional makeup: his remarkable love, grace, justice, and power to conquer death.

Totality astonishes us and greatly exceeds our expectations. Jesus astounds us more. The surprise is so striking that a person cannot understand it unless he experiences it for himself. As Totality immerses us in a simultaneous sunrise and sunset, so the death of Jesus on the cross represents new life for us. This doesn't make sense unless you experience new life for yourself.

As the phases of the Total Solar Eclipse can be mapped to the story of the world, we can also map them to the personal stories of Jesus's followers. Darkness descends on our lives until we encounter Christ, and then his light overcomes the darkness, like Mario's near-death experience. Our personal stories can't be compared to one another, but we can look to each other's stories for encouragement.

Everyone on Earth walks in the valley of the shadow of death.

I know many who abide in the narrow path and look to God in this valley. Their stories speak in strong detail to what the Total Solar Eclipse can only allude to, *No darkness can overcome God's light, but only reveals his beauty more deeply.* These stories beckon listeners to step into the narrow path of faith and look up from there. They inject courage and hope.

In the next chapter, we'll hear more personal stories, and we'll sit together at Mario and Danielle's kitchen table. We'll discover Totality morphs into a second stunning diamond ring, a representation of a new and better covenant between God and people. Together, we'll explore the question that ailed me for year. *Why did Jesus have to die anyway?*

CHAPTER ELEVEN

THIRD CONTACT

The New Covenant

Shortly after I became a homemaker, Jon and I began attending CrossBridge Community Church. Our three children were ages five, three, and six months old. Laurie (the friend who received my first confession in my kitchen) invited us. Jon was not interested in waking up early on Sunday mornings, so I took the kids alone. In those days, he was not open to personal conversations about Jesus and struggled with anger. And me? For the first time, I became acutely aware of how judgmental, selfish, impatient, and arrogant I was. I deeply understood my need for a savior from myself.

"It's nice," I told Jon after a few weeks, "Are you sure you don't want to join us?"

"OK, I'll come," he reluctantly responded, "but don't wake me up until 30 minutes before it's time to go, and if you ever complain on a Sunday morning, I'm out." His childhood Sunday morning memories of church were far from pleasant.

"OK," I said, gladly taking what I could get. I knew Jon and I both needed a savior, and our marriage would likely fail without God's help.

For the next two and half years, I woke Jon up 30 minutes before it was time to go, and we drove 25 minutes to church via North Loop 1604. Many trips were interrupted by Jon's glad voice, "Hey! Let's stop and get Krispy Kreme donuts!" We never left early enough to account for the 15 minutes it would take to stop spontaneously at the donut shop, but I never risked complaining for fear Jon would quit going.

And so, we often arrived 15 to 20 minutes late, walking up the concrete steps to Reagan High School, where our church met. On these steps, I could always count on Dick Freeman, the lead pastor's father, to greet us with a smile. It didn't matter that we were late or that Jon wore a Krispy Kreme hat on his head upon entry. Dick cheerfully welcomed us, and I was thankful to be there.

As time progressed, I volunteered in the Pre-K classes (and somehow arrived early on those Sundays). I couldn't believe "the church people" let a woman like me (a non-church person, whatever that is) serve. Nothing else could have given me a sense of belonging.

When you joined this church, you stood at the front on a Sunday morning, and the pastor presented your family with a bright yellow hard hat. Because our church met at a high school, Sunday setup required many hands. Joining the church symbolized your willingness to work.

"I'd like to join the church. What do you think?" I asked Jon several times.

"No," he always answered, "I'm not interested. Why would we do that? We attend. Isn't that enough?"

I quit asking Jon and started telling God what I wanted. I wanted that bright yellow hard hat. I didn't know what officially joining a church meant, but I wanted to be a legitimate part of God's fellowship and work. After almost three years of attendance,

I became a Sunday school teacher and received an email from our Children's Pastor, Renee. It read, "I noticed you're not a church member; however, to continue teaching Sunday School, you must be one. We have a new member class coming up next month."

I informed Jon, "I'm going to join the church and attend the new member class next month. Want to come with me?"

"Yes," he surprised me.

At the end of the class presentation, Jon and I were assigned to sit at a table alone with Dick Freeman. Dick directed all his attention to my husband. After a lengthy discussion, he asked Jon, "Will you be a little boy for Jesus?" I had never heard anyone ask that question before. I think it was Dick's way of asking, "Will you put your trust in Jesus like a little boy would put his trust in a good Father?"

"Yes, sir," Jon replied.

I had never heard Jon call anyone "sir" in our thirteen and a half years of marriage. My eyes widened in surprise.

Dick then turned toward me, looked into my grateful eyes, and addressed me for the first time that night, "Is *this* the man you want to marry?"

"Yes, sir," I replied.

Turning back to Jon, Dick surprised me, "Jon, do you take this woman, Amy, to be your wife, to have and to hold from this day forward, for better, for worse, for richer, for poorer, in sickness and in health, to love and to cherish…?"

"I do," he said.

Turning to me, Dick asked, "Amy, do you take this man…?"

I did. Dick pronounced us man and wife, and we joined the church.

At that moment, nothing felt different about our marriage. We didn't talk about our renewed wedding vows on the way home, and the door remained closed to talking about Jesus for many months after. Regardless, our renewed vows created a new covenant. With time, conversations about Jesus formed, and we began to serve God together in the children's ministry and on short-term mission trips.

We signed a covenant with the church and renewed our marriage covenant. That's what I see in the next phase of the eclipse: a new covenant.

Third Contact Phase of the Eclipse

Third Contact is the Total Solar Eclipse's second and final Diamond Ring Effect. It indicates the completion of Totality. It means the Moon will move out of the way and no longer stand between the observer and the Sun. This ominous black disc, which, at the end of First Contact, may have brought fear and gloom to the onlooker, has been moving all along. It never had the power to stay in the way forever.

Third Contact is the Total Solar Eclipse's second and final Diamond Ring Effect.

The Moon always moves west to east, but it appears to rise in the east and set in the west because of Earth's rotation. If you could fly into outer space, hover above the North Pole, and look down upon Earth, you would see Earth spinning counterclockwise. You would also see the Moon orbiting Earth in the same counterclockwise direction. Watching keenly, you might notice how the ocean tides follow the Moon. If you're thinking like Newton, you might imagine the Moon exerting an invisible force on the ocean that pulls the waters closer to the Moon. If you're thinking like Einstein, perhaps you almost see the curvature of spacetime causing the water to spill out in the Moon's direction. You notice the Earth spins faster than the Moon revolves around it. In the time it takes the Moon to orbit the Earth, you count a little over 27 Earth spins, or Earth days.

A common question is, "Why don't we get a Total Solar Eclipse every month?" Imagine you're still hovering above the North Pole, looking down upon Earth. You fly away and hover above Earth's equator for a different perspective. You fly away further for a more expansive view of the heavenly bodies. You notice the Moon orbits the Earth in a different plane than the Earth orbits the Sun. Sometimes, when the Moon is between the Earth and Sun, its shadow falls above the Earth. Other times, its shadow falls below Earth. Only when these planes intersect does a Total Solar Eclipse occur.[1]

In Third Contact, the Moon's appointed motion counterclockwise around Earth forces the Moon to inch itself off the Sun's face. The Sun's brilliance bursts forth from one of the Moon's deepest valleys, forming the second diamond ring, another symbol of promise, commitment, and covenant. This time, the solitaire explodes from your left side of the Sun instead of your right.

Third Contact and The Story of the World

Third Contact, or the second Diamond Ring Effect, represents God's second and final covenant with humanity through Jesus Christ. The first Diamond Ring Effect symbolizes God's first covenant with his people, the nation of Israel: *Follow my laws, and I will bless you.*[2] Time and time again, however, his people broke this covenant and didn't follow his beautiful, just, and kind way of doing things (as we saw in Chapter 9). But God had in mind a new and better covenant all along. He told us about it 600 years before Christ through the prophet Jeremiah.

1 Godbole, Ravindra. "Model Guide - Sun Earth and Moon." *YouTube*, YouTube, 24 July 2023, www.youtube.com/watch?v=Y3eaw33Kdp4. Accessed 30 Dec. 2023. (This video provides an excellent moving model of the Sun, Earth, and Moon).

2 Deuteronomy 30:15-18, 1 Samuel 12:14-15

> "The days are coming," declares the Lord, "when I will make a new covenant with the people of Israel and with the people of Judah. It will not be like the covenant I made with their ancestors when I took them by the hand to lead them out of Egypt, because they broke my covenant, though I was a husband to them," declares the Lord.
>
> "This is the covenant I will make with the people of Israel after that time," declares the Lord. "I will put my law in their minds and write it on their hearts. I will be their God, and they will be my people." (Jeremiah 31:31-33)

I love the words, "I will put my law in their minds and write it on their hearts." God can incorporate his law (his good ways) into our beings, making us new people. How? It starts with the crucifixion of Jesus.

God can incorporate his law (his good ways) into our beings, making us new people.

Why Did Jesus Have to Die?

I never understood why Jesus had to die or how it brought me salvation until after we started attending church. I learned God is holy, totally set apart from evil, and perfectly good. Further, God is the only source of life. Without him, there is no life.

And me? I'm unholy without God, mixed with evil (I shared how I learned this about myself in Chapter 8 by asking myself the mighty question: "Why?"). God cannot unite with me in such a state. He cannot join with evil and remain holy.

Sermons don't usually stay in my head for years after I hear them. But I'll never forget when Pastor Kirk stood on stage and described a courtroom scene where we were on trial for our sins (the ways we think and the actions we take that are not in line with God's good and beautiful ways). In this court, God was the judge. We were found guilty. The punishment was death.

Of course, death is the natural consequence. To sin is to separate from God and go our own way, to choose our ways over his. What is death? Death is the opposite of life, separation from the source of life, separation from God. To sin is to separate from God. To separate from God is to die.

But Jesus stood up in court and offered to take our place. The judge accepted, and we were set free. I was able to put myself into the shoes of the person on trial because, for the first time in my life, I knew I was guilty.

It made no sense for the innocent to pay the price just so the guilty could be set free and keep on doing their usual sinful things. Where's the justice in that?

Still, I wrestled with the scenario because it made no sense for the innocent to pay the price just so the guilty could be set free and keep on doing their usual sinful things. Where's the justice in that?

I heard a quiet answer to my wrestling thought: *Amy, what if the guilty party recognizes the love of Jesus and is so thankful to be set free that she no longer goes out to live for herself and according to her own ways but instead lives a new life compelled by the one who loved her and took her place?* For the first time, it made sense.

Understanding what Jesus had done for me compelled me to give my life to him. I no longer tried to do what was right out of fear of harmful consequences. (For example, don't drink too much because it can make you sick, or hide your annoyance and arrogance so people will like you.) Instead, I wanted to do what was right because I loved Jesus and was deeply thankful for what he did for me. I wanted to love people because Jesus loved them. "We love because he first loved us" (1 John 4:19). His ways have become my heart's desire.

A Direct Reversal of the First Sin

This covenant depends not on perfectly following the law but on believing God. "For God so loved the world that he gave his one and only Son, that whoever believes in him shall not perish but have eternal life" (John 3:16). To have eternal life means to be united with the giver of life. Put more simply, the old covenant says, *Follow my laws, and I will bless you, whereas the new covenant says, Believe in my Son, and we'll be together forever.*

God makes a way for us to return to him by providing a direct reversal of the original sin—trust God; trust his love and goodness.

When I first came to know Jesus as my savior from my sin, I thought it strange that all I had to do was believe and trust him. Shouldn't I have to do many good deeds? I asked God about this one day and heard a quiet answer in my spirit: *It's a direct reversal of the first sin.* Adam and Eve distrusted God; they distrusted his love and goodness. God makes a way for us to return to him by providing a direct reversal of the original sin—trust God; trust his love and goodness. It really is that simple. Still, it can be difficult to believe.

Watching the way Jesus walked on Earth helps affirm the truth of it. The story that best helps me to understand involves the two criminals who hung on crosses on either side of Jesus.

Both criminals initially piled contempt on Jesus, as we see in Matthew 27:44 and Mark 15:32. However, we see a change in heart for one criminal in the Gospel of Luke.

> One of the criminals hurled insults at him [Jesus], "Aren't you the Messiah [the Savior]? Save yourself and us!" But the other criminal rebuked him, "Don't you fear God," he said, "since you are under the same sentence? We are punished justly, for we are getting what our deeds deserve. But this man has done nothing wrong."
>
> Then he said, "Jesus, remember me when you come into your kingdom." Jesus answered him, "Truly I tell you, today you will be with me in paradise." (Luke 23:39-43)

One criminal came to a place of repentance. And Jesus, holding no grudge whatsoever, accepted him immediately by responding, "Today, you will be with me in paradise." This story helps us see how quickly Jesus forgives us and how nothing depends on us perfectly following the law. Still, accusing voices come against us constantly, reminding us of our imperfections and trying to keep our heads down in shame.

Attacks on My Identity

You often hear an accuser in your mind if you're anything like me. He's quick to point a finger in your face and remind you that you don't handle everything just right: *You don't manage your calendar correctly. Your house is a mess. You don't eat as you should. You certainly don't treat people with perfect gentleness and kindness.*

About eight years ago, a lady got angry with me about a business transaction. Neither of us acted gently, and she sent me many rude texts afterward. The accusing voice followed me around: *If you were a real Christian (like all those other beautiful women at church), this lady would not be mad at you. She would want to be your best friend. She would even turn to God because of you.* Day after day, this accusing voice bothered me.

Day after day, I lamented my imperfection and defended myself, conversing with my accuser, "I didn't do everything right, but I did my best." I repeatedly replayed my conversations with her, noticing how I could have handled the situation better. My brain was so busy with this dialog that I had trouble sleeping for weeks.

The accusing thoughts came one night at 2 a.m. My husband was out of town, and I was alone in my room. I felt too tired to defend myself or replay the hurtful conversation. Something different rose within me. I spoke out loud to my accuser in a sleepy voice, "You're right about me. I didn't do everything right, and I made mistakes. And you know what? If you go back and look at my whole life instead of focusing on just this one thing, you'll find many more examples like this. And you know what else? I'm going to make more mistakes in the future. So go ahead with the evidence you're collecting against me, and add these to your list. The list is long. Get all this evidence together and take me to court to settle this. I'm done talking with you about it—you're not the judge, so let's take it to the judge and see what he says. I'm not afraid of my judge because he's my father (John 1:12-13), and I adore him. I will gladly submit to whatever the judge says and whatever consequence arises from my actions. And oh, by the way, I will not be representing myself in court. Jesus, my brother, is my advocate. And further? He has already paid the price for my mistakes, so I expect you will lose."

Immediately, the accusing voice fled from my presence, and I slept in peace. The faultfinding voice never bothered me about that particular situation again and I settled into the privilege of John 1:12-13, "Yet to all who did receive him [Jesus], to those who believe in his name, he gave the right to become children of God—children born not of natural descent, nor of human decision or a husband's will, but born of God." Still, the accusing voice periodically tries to make me distrust the incredible gift of daughtership.

Continued Attacks on My Identity

Have you ever felt ashamed for being scared? As if it were a sin? I have. I struggled with fear over a disturbing blood report of someone I love, and I worked hard to combat it by bringing my worries over and over to God and preaching to myself until peace came to my heart.

Have you ever felt ashamed for being scared?

I confessed my struggle with fear to others. *Confessed* as if it was a sin to battle fear. Whenever fear commanded me, "Be scared out of your wits!" an accusing voice joined in the battle: *If you were really a child of God, you wouldn't be afraid; you would have more confidence in your God than that.*

I wondered alongside the accusing voice, *Maybe I don't really know God. Maybe I'm not really a Christian because I'm afraid. I'm terrified.* I wondered alongside this condemning voice until I thought of Jesus in the wilderness. Jesus fasted 40 days and 40 nights there, and he was hungry. The tempter came to him and said, "If you really are the Son of God, turn these stones to bread" (Matthew 4:3).

I had pondered this passage before and always imagined the devil in physical, villainous form, sliding up close to Jesus, hissing in a vicious voice. But I imagined it differently this time. Perhaps the tempter, invisible to man's physical eyes, tempted Christ in the form of thoughts alone, and maybe those thoughts didn't hiss at all. *If you really were the Son of God, you wouldn't be hungry. Undoubtedly, the creator and sustainer of everyone and everything wouldn't be empty and famished in a wilderness. You don't look like the Son of God.* Accusations attacking Christ's identity ran through his mind, yet he remained firm in the truth of who he was.

Similarly, our battles with fear and accusations do not nullify our identity as Christians, beloved children of God. I penned a parable to reveal the invisible battle that raged in my mind and to solidify my grasp on the truth of my identity.

The Parable of the Two Hooded Bullies

King Lynn and Queen Anna sent their daughter Amy to a small hostile country as an ambassador. Shortly after settling into her new place, she heard a loud and unexpected knock at her front door. Hesitantly, she opened the door a little and peered through the crack. Two hooded bullies stood on her porch. The first one, tall and lanky, cried out, "Be scared out of your wits!"

Stunned, she felt her heart race. The one on the left, short and stocky, cried out next, "Look! She panics! She's not the daughter of King Lynn and Queen Anna! They are the most powerful monarchs of all and fear no one!"

Amy quickly shut the door and called her parents. They assured her she did the right thing and encouraged her to continue in that vein.

The scenario repeated for ten days: the duo knocked on the door; Amy refused to let them enter and called her parents, who reassured her. After ten days, the duo got bored with their game and departed.

Indeed, I tell you, King Lynn and Queen Anna were incredibly proud of their daughter, and they took great delight in watching her courage and bravery grow.

So it is with you and God. God is eager to help us through our fears, trials, and struggles. My new friends Mario and Danielle remind me of this.

My Imperfect Obedience and Danielle's Gracious Response

I met Mario, the man who wrote about his near-death experience, at a multi-church event, where he introduced me to his gorgeous wife, Danielle. She and I exchanged numbers. A few weeks later, we chatted briefly on the phone when she agreed I could interview them and share their story in this book you hold in your hands. Two years passed, and I never set up the interview. My father had a massive stroke, my oldest daughter got married, and writing this book went much slower than I had hoped. After two years, I felt ashamed to reach out to Danielle, but I also felt stuck and unable to write. Every time I prayed, I felt God say, *Call Danielle.*

And every time, I retorted, "Oh, Lord, I can't. It's been two years. I'm ashamed to call her now." Strangely, I called him "Lord" yet did not do as he said. I remained stuck and unable to write. Day after day, I prayed and heard him say her name. One morning, the absurdity of my disobedience shone forth like the Sun, and I reached out to her.

On Thursday, October 5, at 10:52 a.m. I texted, "Hi, Danielle! It's been two years since we have spoken! I'm still writing this book. I was ashamed to reach out because I have been a terrible correspondent, but I'm afraid the Lord will not let me go until I do. I would love to meet if you're able and willing."

As the stunning diamond ring surprised me when I thought the Moon would snuff out the Sun in First Contact, Danielle sent the most gracious response the same day.

On Thursday, October 5, at 3:33 p.m., she texted, "Hi Amy! I'm so glad you reached out! I often wondered about you. Yes, feel free to call us. It would be ideal if you could come over for a visit."

Eight days later, I sat at their kitchen table.

Mario and Danielle's Kitchen Table

I rang the doorbell at Mario and Danielle's home. I had never been to their house before and felt nervous about intruding on their day, but the moment I entered their presence, a calm came over me. Danielle welcomed me with a warm smile and hug and invited me to sit with them at their kitchen table.

I asked them, "How have you experienced God in the valley of the shadow of death?" I expected more insight into the story they shared in their book, *How I Know: A Story to Strengthen Your Faith.*

Mario surprised me by sharing a new story, "I have come to an understanding of our role as the bride of Christ."

"What do you mean?" I asked.

Mario proceeded to share about a time after his near-death experiences when he struggled with viewing inappropriate images on the internet. He described it as a struggle for most men in our society.

I must pause and point out that it's a struggle for women, too. I have had struggles with it in the distant past. Several women have confessed this struggle to me. One woman asked, "Amy, why do churches offer classes to help men with this problem but not women? Am I the only woman with this struggle?" You're not.

"As I sat on the couch," Mario began to tell a story, "I felt temptation come over me again. My feet were about to make the move to my iPad. I never used my phone for this activity because I read the Bible on my phone and wanted to keep it separate. I prayed, 'Jesus, please help me overcome this temptation!' Suddenly, love came upon me so strongly that it overwhelmed me. I had never felt so much love in my life. I experienced Jesus as my husband."

Wow! Jesus does not want to punish us. He is the God who moves on our behalf and desires to help us. He overwhelms us with his love.

It's strange for a man to think of God as his husband, so

Mario researched Scripture to ensure his experience matched God's Word. It did. Using his phone, he shared the verses he had discovered and bookmarked through his research. He spouted Scripture like bullets from an automatic rifle.

He is the God who moves on our behalf and desires to help us. He overwhelms us with his love.

- Isaiah 54:5: "For your Maker is your husband—the Lord Almighty is his name—the Holy One of Israel is your Redeemer; he is called the God of all the earth."
- Hosea 2:19 (NLT): "I will make you my wife forever, showing you righteousness and justice, unfailing love and compassion."
- Ephesians 3:19 (NLT): "May you experience the love of Christ, though it is too great to understand fully. Then you will be made complete with all the fullness of life and power that comes from God."

The bullets continued beyond the list I shared here. When Mario came to a pause, I spoke, "Mario, you've shared with me how you've known God under the shadow of great temptation. But surely you cannot wish me to write about that."

"Oh, yes, Amy, I do!" he surprised me. "I'm cured and want to write a whole book about it! I believed the lie that it was just pixels on the screen, not real. But something spiritual is happening when you see it. It's building a wall between you and God without you ever knowing it. It prevents you from speaking to people when you feel led. You feel led to speak to someone and then think: *I can't say anything to that person because I know what I did today*. It leads to shame, which prevents you from doing God's will and purpose in your life."

I can relate. I feel shame whenever I miss the mark, lose patience, get annoyed, behave rudely, or entertain ungodly thoughts. I hear a voice saying, "If you really were a Christian, you wouldn't have done that. You're not a legitimate Christian, so keep your head down and step back. Let the real Christians get involved in God's work." But our job is to follow Jesus. When we get off track, we simply turn to him again. As Mario turned to God, God overwhelmed him with his love, which gave him the power to overcome.

So...What's the Point?

The point is that when you trust and love Jesus, you belong to him forever, no matter what you've done, how you feel, or what you struggle with. That's his promise. It's simple and almost sounds too good to be true, especially when you know how unholy you are and how holy God is.

Your imperfections are like the craters of the Moon; they make space for God's goodness, love, and light to shine through and excite the world like the Diamond Ring Effect of the Total Solar Eclipse. Your mistakes, wounds, and scars are not to be hidden or become a source of shame. No, when we align our imperfect, messed-up selves with God, he surprises us with a stunning and beautiful testimony of his goodness, power, and love. He's so good and loves us so much that he uses his power to make beautiful light out of our darkness.

Third Contact's second Diamond Ring Effect of the eclipse testifies to God's second and final covenant with humanity. The first covenant depended upon people following God's law perfectly; through it, we learned no one could do it in their own strength. But God responded to our weakness with the second covenant, which is about trusting Jesus. My friend Brittany said, "We are not called to follow perfectly. We are called to follow the perfect one." When we get off track, like I did driving across Texas, we can simply turn around and follow him again, trusting that he is waiting for us with open arms like the prodigal son's father (Luke 15:11-24). As we follow Jesus, we discover he transforms us day by day, pushing out the darkness in our hearts and flooding us with the light of his love.

We are not called to follow perfectly. We are called to follow the perfect one.

In this next chapter, we explore the last phase of the Total Solar Eclipse and discover how it parallels the current phase of humanity in which we now reside. We'll explore several questions. *What is God doing these days? How are God's people involved? How can I forgive someone who hurt me? How can I know if I'm being deceived and walk in truth?*

CHAPTER TWELVE

FOURTH CONTACT

Restoration

Have you ever regretted something you did long ago, maybe even decades ago, so much so that you grieve and mourn it in the present? I have. Late in the evening, when I was a mother to two girls in high school and a boy in middle school, I deeply regretted a situation I put myself in over twenty years ago. I sat in my bedroom with the door closed and wept bitterly. My son Freedom heard me and entered the room.

"Mom, are you okay? Why are you crying?"

I couldn't explain to my young son why I cried so deeply—it would have been inappropriate. "Everything's fine, son. There are no problems. I'm only very sad and can't tell you why."

"Well, if you can't tell me, then you have to tell Jesus, and I'm not leaving this room until you promise me you will talk to Jesus about it," he commanded seriously.

"I'll do it," I assured him with a weak smile through my tears.

Satisfied, he hugged me and left the room.

I honored his request and began to talk to Jesus about it. "Oh,

Jesus, where were you when I was in that painful place so long ago? I deeply regret it today!"

Immediately, in my mind's eye, I could see myself in that place over 20 years ago. I was not proud of myself. In that place, I could see Jesus in the room with me, the room I should not have been in, committing the sin I should not have been involved in. I wept all the more bitterly. "Oh my God, you were there, and you saw me? And you died on the cross for *that?*" I felt like throwing up and wept all the more as remorse and hurt overwhelmed me.

> I see God's beautiful heart to separate us from sin and restore us back into himself.

I heard a loving and unexpected response in my spirit: *Amy, no, I did not die for that. I died for you. And yes, I saw you in that place, and when I did, I could see you separate from what you were doing. And look! Look how separated you are now! You are so separated that the thought of it grieves you deeply today.* Comforted, I felt God's pleasure as he smiled at me. As I write this, I'm reminded of Jesus's words, "Blessed are those who mourn, for they will be comforted" (Matthew 5:4). I see God's beautiful heart to separate us from sin and restore us back unto himself in the Fourth Contact phase of the eclipse.

Fourth Contact Phase of the Eclipse

The Fourth Contact phase of the eclipse is like the First Contact phase in reverse. Rather than a growing partial eclipse, it's a shrinking partial eclipse and typically lasts about an hour and a half. You see the same crescents of light created on the ground under the trees, but they face the opposite direction this time.

I wonder—if you were at a high enough vantage point with a

landscape view, could you see this tremendous shadow retreating at the beginning of Fourth Contact?[1] During the 2017 eclipse, I didn't think about the formidable, 150-mile-wide shadow traveling across Earth at over 1,500 miles per hour.[2] Being nestled in the trees with no view of the landscape, I wouldn't have been able to see it coming or going anyway.

During Fourth Contact, the black disc slowly slides off the face of the sun. You must put your protective eclipse glasses back on to look up and see the growing crescent in the sky. In 2017, we lost interest once we experienced First Contact, Second Contact (the first Diamond Ring Effect), Totality, and Third Contact (the second Diamond Ring Effect). No one cared to put on paper glasses and look up at the sun once Fourth Contact gained momentum. We knew the Moon would continue its journey in front of the Sun, and the Sun's full brilliance would be restored to us once more. *Restored* is the word I see in Fourth Contact.

We knew the Moon would continue its journey in front of the Sun, and the Sun's full brilliance would be restored to us once more.

Fourth Contact and the Story of the World

Connecting the story of the world to the phases of the Total Solar Eclipse, humanity now resides in Fourth Contact, the last stage. We mapped the first phase to the Fall when darkness appears to overtake light. In Fourth Contact, light pushes away the dark until

1 Kentrianakis, Mike. "Solar Eclipse | Incredible Footage from Flight." *YouTube,* ABC News, 10 Mar. 2016, www.youtube.com/watch?v=hMvTZaqMDx8. Access 18 Jan. 2024. (In this video, you see the shadow encroaching from the high vantage point of an airplane)

2 Wright, Ernie, and David Ladd. "NASA Scientific Visualization Studio." *NASA,* NASA, 18 Dec. 2023, svs.gsfc.nasa.gov/5186/. Accessed 30 Dec. 2023.

it's completely gone. This plays out through the church—not a physical building where people meet on Sunday—but in the lives of Jesus's followers, who, together, look and walk like Jesus did. Jesus's followers differ from most people; they deeply love others, seek ways to help them, and speak life into their broken areas.

Jesus taught his followers about their role in the world when he said, "You are the light of the world. A town built on a hill cannot be hidden. Neither do people light a lamp and put it under a bowl. Instead, they put it on its stand, and it gives light to everyone in the house" (Matthew 5:14-15).

We spread light by asking God to include us in his work to push out darkness.

Jesus's followers are like candles lit by Christ. We spread light by asking God to include us in his work to push out darkness. For almost two decades, my pastor has taught us to pray what he calls The Include Me Prayer. It's simple: "Lord, I love you. Will you please include me in what you're doing today?"

And what is Jesus doing every day? He's pursuing people, taking them "out of darkness into his wonderful light" (1 Peter 2:9). At a conference I attended about a decade ago, Andy Reese, co-author of the book *Freedom Tools,*[3] showed us God's heart for people by highlighting four biblical parables: the parable of the lost sheep, the lost coin, the prodigal son, and the angry brother. When I think about how God spreads his kingdom of light in this phase of humanity, these four parables and Andy's teaching on them come to mind.

The Lost Sheep

Jesus told a parable about the lost sheep to show us God's heart for people. The shepherd left the 99 to go after one lost sheep. And

3 Reese, Andy, and Jennifer Barnett. *Freedom Tools:* For Overcoming Life's Tough Problems. Chosen Books, 2015.

when he found it, he joyfully put it on his shoulders, brought it home, and then threw a party to celebrate the rescue (Luke 15:3-7).

I was once a lost sheep. When I was a teenager drinking bootlegger alcohol, I didn't feel rebellious. I was simply wandering around with no intention. I gave no thought to my purpose and sometimes found myself entangled in ungodly circumstances. I roamed around thoughtlessly for years until something changed.

Jesus found me, picked me up, and brought me home, primarily through my neighbor and friend Laurie. Laurie led a Bible study in her home, received my confession with grace and love, prayed for me, and invited me to church. I will love her forever because of it. When I put myself in this narrative today, I relate to one of the 99 sheep. What I want to do is follow my shepherd.

How do Jesus's followers push back the darkness in this phase of humanity? They pray the Include Me Prayer: "Dear Lord, I love you. I want to follow you wherever you go. Will you please include me in your rescue work today?" Then, they follow Jesus in his rescue of his lost sheep, becoming the hands and feet of his shepherd's heart. Jesus's followers introduce lost sheep to the Good Shepherd, who can bring them home. When Jesus does bring them home, his followers celebrate with him, protecting the celebration from any hints of shame.

The Lost Coin

Jesus told a story about the lost coin to show his heart for people. A woman had 10 coins, and she lost one. She lit a lamp, swept the house, and searched carefully until she found it. When she found it, she threw a party to celebrate (Luke 15:8-10).

A coin doesn't wander away; it gets displaced by someone else. You may have forgotten your value because someone else has not recognized your worth and put you in a place you do not belong.

When you find yourself lost in a dark, dirty place, may you know Jesus as the one who lights a lamp, sweeps the room, and searches carefully until he finds you. When your worth is covered in dirt, and you have forgotten your value, may you know God as the one whose image you bear. A coin bears the image of the government, maybe Caesar's, so it belongs to Caesar. But you? You bear the image of God because you belong to God.

How do Jesus's followers push back the darkness in this phase of humanity? They pray the Include Me Prayer: "Dear Lord, I love you. I want to be with you as you light up and clean up the dark places. Will you please include me in your rescue work today?" Then, they follow Jesus in his search for the lost coins. His church becomes the hands and feet of his treasuring heart. Jesus's followers light up the dark places, clean out the dirty places, and uncover his treasured jewels. When they are found, Jesus's followers remind them of their worth and celebrate, protecting the celebration from any hints of shame.

Jesus's followers light up the dark places, clean out the dirty places, and uncover his treasured jewels.

The Prodigal Son

Jesus told a story about the prodigal son to show his heart for people. A father had two sons. The younger asked for his inheritance early, and his father gave it to him. Shortly afterward, the younger son took all his money to travel to a faraway country, where he wasted his inheritance on reckless, extravagant living. He ran out of money, the economy went sour, and he had no friends to help him. At his lowest point, he decided to return home, hoping his father would allow him to at least become a servant. But his father saw him while he was still far off, ran to him, fell on his neck, and kissed him. Having rehearsed what he would say to his father, the son said, "Father, I have sinned against heaven and you, and I'm no longer worthy to be called your son. Make me like one of your hired servants."

But the father responded differently than the son imagined. The father turned to his servants and said, "Bring out the best clothes for my son, including a ring for his hand and sandals for his feet! And bring out the best food because we're throwing a party to celebrate his homecoming!" (Luke 15:11-23).

When the prodigal arrives, Jesus's followers run to greet them, shower them with love, remind them of their worth, and celebrate.

There is no god like our God, who loves and longs for people as his own sons and daughters. The only explanation is that God is our good father, and we truly belong to him.

How do Jesus's followers push back the darkness in this phase of humanity? They pray the Include Me Prayer: "Dear Lord, I love you. I want to stand with you and open my arms wide to the returning prodigals like you do. Please include me in your work today to watch for them and welcome them home." Then, they stand with Jesus as he watches for the return of the prodigals. The church becomes the hands and feet of God's father-heart. When the prodigal arrives, Jesus's followers run to greet them, shower them with love, remind them of their worth, and celebrate, protecting the celebration from any hints of shame.

The Angry Brother

Jesus told us a story about the angry brother to show God's heart for people. The prodigal son's older brother (see the previous parable) heard about the celebration his father hosted. The older brother became angry and refused to attend, so his father went out and pleaded with him. Disgruntled, the angry son responded bluntly to his father, "Look! All these years, I've been slaving for you and never disobeyed your orders. Yet you never gave me even a young goat so I could celebrate with my friends. But when this son of yours who has squandered your property with prostitutes comes home, you kill the fattened calf for him!"

"My son," the father said, "You're always with me, and everything I have is yours. But we had to celebrate and be glad because this brother of yours was dead and is alive again; he was lost and is found" (Luke 15:25-32).

God wants to include us in his beautiful work, and he holds no good thing from us.

God's heart of forgiveness is grander than we can imagine, and he loves his children so much that he calls us up to his beautiful standards: to be like him, love like him, and forgive like him. "Everything I have is yours," God says. "This son of mine is your brother," God says. "You and I are meant to throw these parties together," God says. God wants to include us in his beautiful work, and he holds no good thing from us, but sometimes we forget.

When I sat at Danielle and Mario's kitchen table, Danielle shared with me about a time she felt angry with God. "We have this disease of forgetfulness," she explained. She's right. We forget God is good. We forget our unity with him, our oneness with him. We need each other; when one of us forgets, the other can remind us of the truth.

How do Jesus's followers push back the darkness in this phase of humanity? They study the Bible to know God's heart. They pray the Include Me Prayer: "Dear Lord, I love you. I want to remember and to help my brothers and sisters remember that you hold back no good thing from us. Please include me in your work today to help us remember that you are good and we are loved." Then Jesus's followers watch out for each other. And when one of us forgets and starts to believe the lie that God is not good, we reach out to them to help them recognize that lie and exchange it with the truth. Afterward, we return to co-hosting parties with God, protecting those parties from any hints of shame.

Jesus's followers watch out for each other.

Over 2,000 Years Later

Jesus spoke and modeled these parables over 2,000 years ago. Those of us who have been lost sheep, lost coins, prodigal sons, and angry brothers have been rescued and loved by God. For 2,000 years, his love has compelled us to go out into the world and point to Jesus. We follow Jesus in his search for the lost sheep and coins. We earnestly pray, wait, and watch for the prodigal sons to return so that, like God, we can receive them with open arms. When one of us forgets God's goodness, we reach out and remind each other of his love.

No darkness can overcome the light Jesus lit in his followers 2,000 years ago. We keep spreading his light and overcoming the darkness around us with invisible forces we can't see with our eyes. It's like how the Moon moves away from the Sun in Fourth Contact. The invisible force of gravity does it, and we can do nothing to stop it.

My Restoration

After Jon and I joined CrossBridge Community Church, we grew (and continue to grow) in restoration. Day by day (sometimes little by little and other times in leaps and bounds), we are renewed as God's kingdom of light expands in our hearts and minds, pushing out darkness. For me, two areas, in particular, have been catalysts, like bulldozers pushing the Moon out of the Sun's way in Fourth Contact: walking in forgiveness and exchanging lies with truth.

Forgiving

Unforgiveness is like a dark shadow over your life. If you're anything like me, forgiveness can be difficult. You want to forgive. You say you forgive. But then old offenses flood your brain against your will at the most inopportune times; you become angry, and wonder if you really can forgive. A dark bitterness sets in, steals your joy, and hurts your relationships. Unforgiveness makes us slaves to the darkness of deception and division. Forgiveness ushers in the light of truth, freedom, and unity.

Forgiveness ushers in the light of truth, freedom, and unity.

We sometimes need help to walk out forgiveness. I'll never forget the first time I reached out to Jesus's followers to help me forgive someone. I wrote a parable about two sisters and a king to describe what I learned. When I find myself in a state of unforgiveness, I look back on this parable, and it helps me forgive.

A Parable of Forgiveness

Once upon a time, two sisters met at a rich, sturdy table, beautifully cut from one piece of wood called the Table of

Fellowship. There, the older sister Amelia promised her little sister Lily 100 blossoms in an array of colors from the deepest red to the darkest violet. When the time came to deliver, only 50 flowers arrived. The bouquet was lovely indeed, but half the spectrum of colors was missing. The bouquet was filled with only red, orange, and yellow hues. Lily, well-pleased with the 50 flowers, decorated the table with them and eagerly awaited her older sister to deliver 50 more.

They met daily at their table, sharing words, smiles, and food among the fiery hues, but when the days turned to weeks, and Lily never collected the other colors expected, she frowned. As one who kept a careful record of life, she created a debt list.

> Debt List
> My sister owes me:
> 12 flowers in shades of green
> 13 flowers in shades of blue
> 12 flowers in shades of indigo
> 13 flowers in shades of violet

Lily kept the debt list at the Table of Fellowship next to her as a reminder each time they came together, suspecting her sister would take notice and immediately make good on her promise. Amelia didn't.

Frowning, Lily stamped "PAST DUE" on the list in big red letters and discretely slid it across the table, expecting Amelia to notice and pay her debt. Amelia didn't.

When weeks became years, Lily handed it directly to Amelia at the table with stiff arms. Although the debt list compelled Amelia's head to lower so their eyes no longer met, it did not compel her to pay her debt.

As the years turned to decades, Lily grew in wisdom. She understood Amelia could not pay what she did not have. Amelia paid what she could, and, indeed, her flowers were lovely and good. The debt list only lowered Amelia's head and made Lily slave to a frown. Lily missed fellowship with her sister. She came to deeply appreciate the fiery flowers Amelia offered. Despising the debt, Lily took it off the table and locked it away in a dark drawer. She wanted to forgive and forget, tired of her discontent and failed attempts as a debt collector. However, this list was magical.

On one of her happiest days, when Lily smiled and danced among the red, orange, and yellow glow, setting the table for a feast with her sister, the debt list suddenly appeared with its big red stamp.

"Go away!" Lily cried. "I locked you up!" However, the debt list, alive with its own movement and voice, cried aloud,

It's only right.
It's what is due.
You must collect the debt or sue.
Don't doubt my song.
You know it's true.
You're incomplete while it's past due!

Frowning, Lily locked it away again, and this time, she put it under a thick book to muffle its song and keep it in its place. But once again, on one of her happiest of days, the debt list found its way to her at the Table of Fellowship and sang,

It's only right.
It's what is due.
You must collect the debt or sue.
Don't doubt my song.
You know it's true.
You're incomplete while it's past due!

Frowning, Lily tore it to pieces and threw it in the trash, sure that would be the death of it, but the list didn't die. It made its way back to the Table of Fellowship again singing its familiar demanding song.

Desperate, Lily confided in another, "What can I do?"

"Oh, you can never get rid of a magical list like that on your own. It's too powerful," spoke the other.

Intent on finding freedom from the incessant and demanding list, she gathered a group of her friends, the most powerful and wise advisors in the land.

"Take the list out of the drawer," they directed.

She did. With eyes kind, steady, and right, they examined it.

"Oh yes, the debt sings partial truth," her friends, the advisors, said. "These flowers are owed to you. Take it to our Father King. We will come with you."

With advisors by her side, Lily held the list in both hands as she approached the King of kings. Head down in reverence, hands full of debt, both arms extended in hope, she submitted the list to him on bended knee.

Gathering the note in his hands, the King spoke, "My daughter, I am so rich."

His deep, love-voice endowed her with courage, lifting her eyes from his hands to his face, and Lily watched as his compassionate eyes and serious lips continued, "All I have is yours. Being so rich, you must not spend yourself in the act of collecting debts."

As he spoke, his right hand swept in front of him and to his side to reveal a field of flowers that extended farther than Lily's eye could see in paints, infinite and unimaginable. "Pick whatever you'd like, my daughter. My supply is limitless, and all are complete in me."

Lily returned to the Table of Fellowship, free from the dark, demanding debt list and full to overflowing with light, forgiveness, and flowers.

Practical Steps to Forgive

If we need to forgive someone, it's because they owe us something. Perhaps your boss owed you recognition or approachability. Your father owed you protection, provision, or an encouraging word. But maybe the word "owed" feels too strong because you know your father did his best. You can use the word "wish." You wish your mother had given you comfort, wisdom, and guidance. You wish your spouse had given you love and respect.

If we need to forgive someone, it's because they owe us something.

The first step in forgiveness is to count up the debt, that is, to list what the other person owed you or what you wish they gave you.

The second step is to cancel the debt, that is, to proclaim that you will not strive to collect the debt from that person. Instead, you will look to God for everything you need and wish for, such as recognition, approachability, protection, provision, comfort, guidance, etc. In the parable, Lily counted and canceled the debt.

The third step is to give your debt list to the Lord. Lily gave the list to her King (Jesus) and found that he would provide everything she needed, wanted, and more.

In the forgiveness process, Lily's eyes were also opened to a fresh perspective of the one who owed her. She realized she had judged her sister harshly. Her sister Amelia had done her best and brought all the flowers she could to the Table of Fellowship. We can ask Jesus for this fresh perspective of the person who owes us, and finally, we can bless that person and pray for them.

It's worth noting that forgiveness doesn't always bring the kind of unity seen in this parable. Sometimes we need to forgive someone that we do not need to unite with (for example, we do not need to unite with an abusive person). Sometimes we need to forgive someone who has passed away. In these cases, forgiveness will not usher in unity with the person we have forgiven, but it will usher in unity with God, who freely forgives us (Matthew 6:14-15, Matthew 8:21-25).

Unforgiveness can also place us under a shadow of deception. Lily came to believe the lie that she was incomplete until her sister provided everything she wanted. Forgiveness ushered in the light of truth—no one can complete us, except God. And that leads us to the next catalyst that has played an important role in pushing out the darkness in my heart: exchanging lies for truth.

Exchanging Lies with Truth

Exchanging lies for truth is another way we push away darkness. Lies are bondage makers, primarily undermining the goodness of God and his great love for us. Lies make us slaves to fear and keep us from the light of God and our purpose. Truth is the bondage breaker.

Exchanging lies for truth is another way we push away darkness.

About five years ago, I received an email from a friend, full of encouraging words, including these four, "You are a dreamer." She couldn't have known I had entered a new season of dreaming and setting goals to achieve that dream. Every time I took steps toward the goals, a question came to mind with an accusing tone, "Who do you think you are?" Although my friend didn't use the word "just" in her email to me, my accuser did and placed her words alongside his, "Who do you think you are? You are just a dreamer. You're too little and silly to achieve anything." These words made me believe I was incapable of reaching for the dream in my heart. The accusing words were like an eclipse, blacking out the plans I made to attain my goals.

As I wrestled with feeling too small and too incompetent to move forward, I recognized the lie and replaced it with the truth that God is my good and powerful father who planted a purposeful dream in my heart and had every intention of giving me everything I needed to bring it to fruition. To remember God's goodness and to make visible the invisible forces that worked against me, I wrote a parable about it.

A Parable about Replacing Lies with Truth

One day, while I dared to dream, a tall, slender man in a stylish suit stepped toward me with a purposeful stride and a charming face. Without greeting me, he asked, "Who do you think you are?"

Seeing my response was not immediately forthcoming, he took a small notepad and pen from the inside of his coat pocket and began scratching words with his left hand while he spoke with tight lips, pausing to peer at me between each sentence as though painting a portrait. He spoke what he wrote:

"You are just a dreamer.

You are just a peacemaker.

You are just a mother.

You are just a wife."

"Wait," I interjected awkwardly in an attempt to allow burning truth to break forth from my bumbling body, "I am also a child of God."

Cocking his head to the side, he obliged, mumbled, and scratched, "You are just a child of God."

With an important air and exaggerated arm, he tore the paper from his pad and folded the paper into a perfect pill with a swiftness that comes only from plentiful practice. "Take this," he smiled kindly as he presented this pill like a gracious gift.

Seeing my hesitation, his other hand offered a tall, clear glass of water to wash it down. "To remember yourself rightly, lowly as you are," he said. He bowed shallowly as he spoke and waited patiently in his paltry bend with the pill held out in his hand at the perfect height for my bite.

Oh, to remember my smallness—it's truth I surely must eat, I thought as I took the tablet with a lowering head. Its bitterness burned on my tongue, down my throat, and in my belly. I reached for the water only to discover the man no longer offered it. Furthermore, I had shrunk very small!

The strange, tall man now had his hands on his hips. He leaned down close to laugh loudly in my face. Then, rigging my dream on the end of a stick, he dangled it high above me. I leaped, spun, and lunged to seize it. Impossible! He taunted until he saw I had given up, and then, leaving my dream high on a shelf to tease me, he departed.

I craned my neck to see my dream, high and large above me. Finally, I decided it was probably best not to look up at it, for it strains the neck of someone so small.

Just a dreamer.

Just a peacemaker.

Just a mother.

Just a wife.

Just a child of God. The bitter words swaggered in my mind, reducing my identity to its most minimal level. Out of habit, I reached for my Light and Book, always tucked in the folds of my clothes, with me wherever I go, and I read.

Oh, he's lying! I pulled out my journal and pen. With my left hand, I spoke what I wrote:

I am a dreamer!

I am a peacemaker!

I am a mother!

I am a wife!

I am a child of God!

I am a child of God, and not just in the titular sense, but in the total sense! A legitimate heir! See? The accuser almost had it right, just one dastardly blunder: Make "just" a noun. *Use a comma like this: I am just, one who lives by faith.* Now it's stated just right, and is it not just glorious?

Word washed down like water. With a lifted head, I grew to my rightful size, plenty big enough to treasure a dream that fit well in my hands.

Practical Steps to Exchange Lies with Truth

The first step to exchanging lies with truth is recognizing we're believing a lie. We don't usually believe lies on purpose. In the parable, the liar took the truth and altered it just enough to wrap it in a whole lot of falsehood. Lies are deceitful. So how can we know when we are believing one?

The truth of Jesus is convicting, encouraging, and loving.

My friend Jill pointed out that lies are condemning, accusatory, and shaming. The truth of Jesus is convicting, encouraging, and loving. We can look again at the four parables Jesus told about the lost sheep, the lost coin, the prodigal son, and the angry brother to recognize the voice of Jesus, the voice of Truth.

If you hear an idea in your head, and it sounds harsh, critical, or speaks to your worth in a negative way, it's a lie. It's not God's idea. Lies lower our heads in shame and make us feel like little, incapable, unwanted stepchildren to God.

When my children have a dream, I'm all in. Want to play the piano? I'll provide piano lessons. Have it in your heart to sing? I'll provide voice lessons. Want to do gymnastics? I've got you covered. I often tell them, "I'm behind you 100%. Whatever you need, I'm on it."

But I'm prone to believe the lie that God does not love and adore me like that because I couldn't be a true daughter of the creator of the universe. It's easy for me to believe the lie that if I ask him for bread, he may give me a stone because I'm not *really* his, and he's not *really* as good as the Bible says. Lies separate us from God, and they also separate us from our God-given purpose.

This parable directly relates to you because the dream referenced in the parable is the dream to write this book you hold in your hands. If I believed the lie that I was unfit to write this book, you'd be doing something else right now because this book wouldn't exist. Lies have a motive, and knowing their motive helps us uncover them. Their motive is to keep you and me separated from God, each other, and our purpose.

If an idea keeps us from reaching for our God-given dreams, it's a lie. How can we know what's true? We can ask Jesus. We can test an idea against the stories of Jesus shared in the Bible. The truth of Jesus is loving, encouraging, compassionate, and kind. Jesus is the knower of our worth and the lifter of our heads. He is the great encourager and the best teacher. Jesus may convict us of wrong behavior, but only to free us from it and draw us closer to him and what he's doing (the parable of the angry brother). Jesus will never shame us into isolation. His heart is to pull us close, keep us close, and celebrate our togetherness. When we look at the four parables (the lost sheep, the lost coin, the prodigal son, and the angry brother) we find this is true about him.

If an idea keeps us from reaching for our God-given dreams, it's a lie.

How do we exchange lies for truth? We reject the lie and embrace the truth. We follow through with our behavior. We write the book. God's light spreads, and darkness is pushed away.

So...What's the Point?

The point is we're living in Fourth Contact, the last phase of this Earth, when God's light spreads and pushes away the darkness through Jesus's followers. While First Contact represents our separation from God through our union with sin, Fourth Contact represents God's work to separate us from our sin and then reconcile us to himself. This separation from sin is called freedom, and it's part of God's plan to restore all creation to its good beginning.

This separation from sin is called freedom, and it's part of God's plan to restore all creation to its good beginning.

In the parable of the lost sheep, the lost coin, the prodigal son, and the angry brother, we find God's heart to rescue people from the darkness of isolation. We can pray to be included in God's plan to usher in the light of restoration and togetherness.

We can walk in the light of forgiveness with God's help by counting up the debt owed to us (or the debt we wish was given to us) and canceling the debt, turning it over to God. Finally, we can look to God to complete us.

We can walk in the light of truth. We can differentiate between lies and truth. Lies condemn, accuse, and shame. Lies lower our heads in shame and breed division, anger, and emptiness. Truth convicts, encourages, and loves. Truth lifts our heads and brings unity, peace, and strength. Truth aligns with Jesus and his way of doing things, which we can know through the Bible, his followers, and the still small voice inside us, his Spirit.

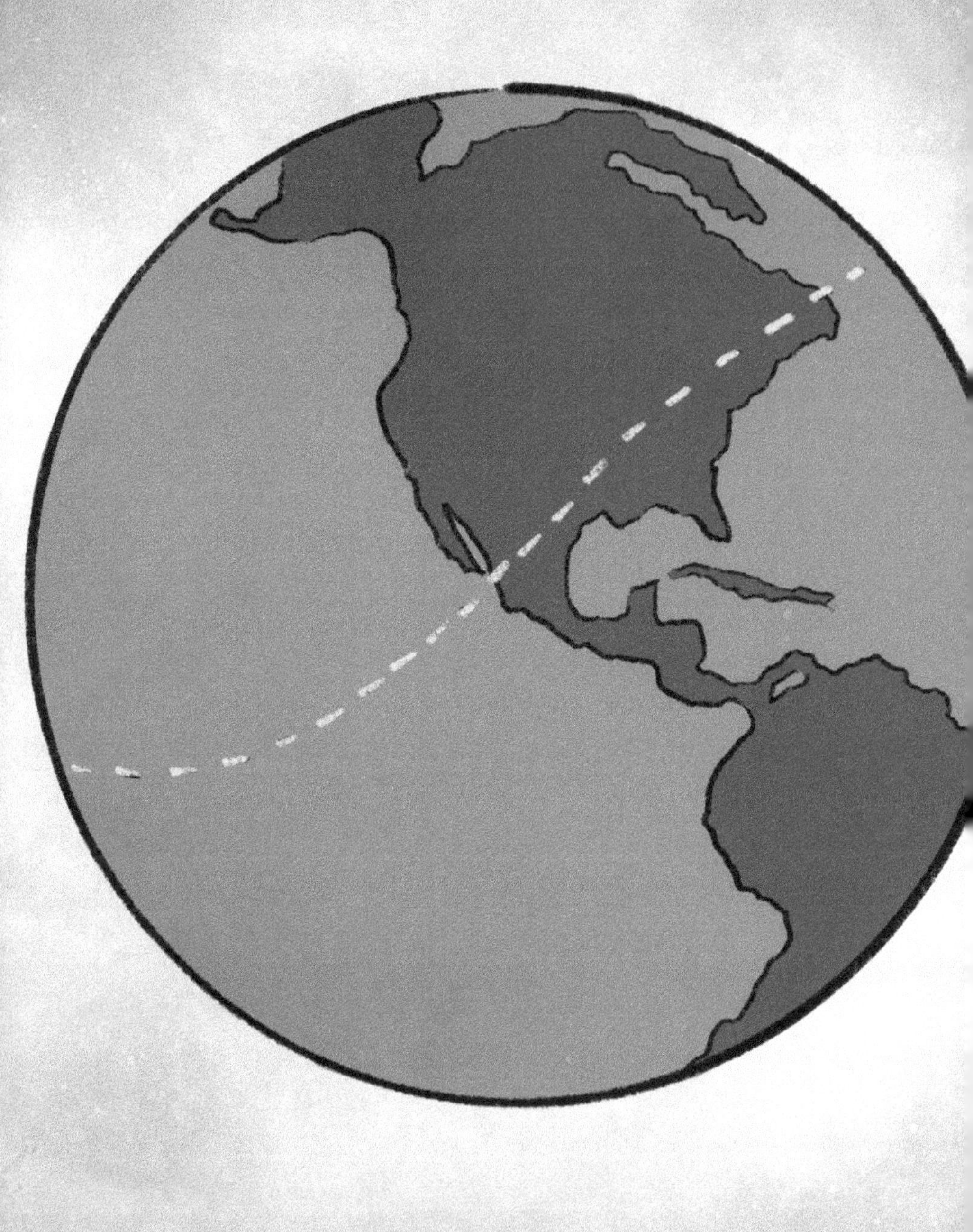

CHAPTER THIRTEEN

THE NARROW PATH

You're Invited

We were newlyweds when Jon and I drove from the Texas Panhandle to the Gulf Coast. The trip was over 500 miles and would take all day. In 1996, we didn't have smartphones with navigation apps, so we studied a Texas map, planned and memorized our route, and then went on our way. Near the beginning of the trip, I drove, and Jon napped. He slept for over an hour while I happily daydreamed, enjoying the view of the Texas plains where you can see as far as your eyes will let you because no trees and no hills stand in the way.

The first words out of his mouth when he woke up surprised me, "Where are you going?"

"To your parents' house in Corpus Christi, of course."

"No, you're not. You're going in the wrong direction. What road is this?" he asked as he unfolded the map.

I wondered how he could be so confident. *What does he recognize that I don't? Surely, I'm not going the wrong way.* I scanned the right side of the highway for a road sign. "I'm on Interstate 20."

"Amy, you were supposed to turn off back there in Sweetwater," he laughed. "We need to turn around."

Going the wrong direction for a little while didn't matter much because we still made it to his parents' house that day. But to get where you want to go and see what or who you want to see, you must pay attention and get on the correct route. So it is with the path of Totality and a Total Solar Eclipse. There's only one path to experience a total eclipse of the Sun, and you want to ensure you're on it so you don't miss the most glorious light show known to humanity.

To get where you want to go and see what or who you want to see, you must pay attention and get on the correct route.

The Path of Totality

When the Moon totally eclipses the Sun, it casts a fuzzy shadow on Earth. If you look through the window of an airplane or the International Space Station during a Total Solar Eclipse, you would see the Moon's round shadow on the Earth. This shadow is composed of two parts. The center of the shadow is dark and called the *umbra,* the Latin word for "shade, shadow."[1] The outer part of the shadow is called the penumbra, a combination of the Latin words *paene* and *umbra,* where *paene* means *almost.*[2] This almost-shadow is so large that it nearly covered the entire continent of North America in 2017.[3] The umbra, on the other

1 "Umbra Definition & Meaning." *Merriam-Webster,* Merriam-Webster, www.merriam-webster.com/dictionary/umbra. Accessed 2 Nov. 2023.

2 "Penumbra Definition & Meaning." *Merriam-Webster,* Merriam-Webster, www.merriam-webster.com/dictionary/penumbra. Accessed 2 Nov. 2023.

3 Wright, Ernie. "NASA Scientific Visualization Studio." *NASA,* NASA, 24 Oct. 2023, svs.gsfc.nasa.gov/4314. Accessed 5 Nov. 2023.

hand, is small. In 2017, the umbra spanned approximately 71 miles.[4]

To stand in the shadow of the umbra is to experience Totality. This may be one of the most important aspects of a Total Solar Eclipse to be aware of. A total eclipse of the Sun occurs at a specific date and time and at a specific location. What may be a Total Solar Eclipse for me at 1:32 p.m., on April 8, 2024, on the north side of San Antonio, Texas, will only be a partial eclipse for you in downtown San Antonio. In other words, if you're downtown or south of town, and I'm on the north side, I'll experience the Diamond Ring Effects of Second and Third Contact with Totality in between. You, on the other hand, will know no diamond rings and no corona. If you have eclipse glasses, there will never be a safe time to take them off and look at the Sun during the eclipse. You will not experience Totality. I repeat: You will not experience Totality.

To stand in the shadow of the umbra is to experience Totality.

The Moon's umbra will never cross downtown San Antonio in 2024. You must get in the path of Totality, under the umbra of the Moon, to experience the Total Solar Eclipse. One small step outside the path, outside the umbra, and you fail to experience the diamond rings and the fiery crown. What a shame it would be to be so close to the most majestic light show known to humanity and miss it!

The closer you get to the middle of the path, the umbra's center, the longer you experience Totality. For example, on April 8, 2024, I will experience two minutes and 42 seconds of Totality if I'm in Helotes, Texas. However, if I'm at Kerrville-Schreiner Park, approximately 60 miles north of Helotes, I will experience over four minutes and 25 seconds of Totality. If I'm in downtown San

4 Scharf, Caleb. "Eclipse: It's All about the Umbra." *Scientific American Blog Network,* Scientific American, 18 Aug. 2017, blogs.scientificamerican.com/life-unbounded/eclipse-its-all-about-the-umbra/. Accessed 5 Nov 2023.

Antonio, about 26 miles south of Helotes, I will not experience Totality.

So, how do you know where the path of Totality will be? You refer to an eclipse map; many are provided for free online.[5,6,7] The 2024 Total Solar Eclipse will sweep across 15 states from Texas to Maine.[8] You want to get in the middle of the path of Totality where, in some places, you may experience over four minutes of Totality (time to enjoy the pearly kinetic crown). If you are near the path's edge, you may only experience a few seconds in Totality. The San Antonio International Airport is on the edge of the path of the 2024 eclipse. Stand on a precise point of the runway to experience 10 seconds of Totality. Stand on the wrong point of the runway, and you experience zero.[9]

As the Sun, Earth, and Moon perfectly align to create a Total Solar Eclipse, we observers must precisely align ourselves to experience Totality. We don't get to command the Earth, Moon, and Sun to align with us. No, we must align with the heavenly authorities.

In 2017, we drove almost 1,000 miles to align ourselves under the center of the umbra of the Moon and experience Totality from Fort Campbell, Kentucky. Our best friends experienced the same eclipse outside the path of Totality from San Antonio, Texas. For us, it was awe-inspiring and spoke to our bones. For our friends outside the umbra, it was simply a neat experience, a partial

5 Jubier, Xavier. *Mexico - USA - 2024 April 8 Total Solar Eclipse - Interactive Google Map - Xavier Jubier,* xjubier.free.fr/en/site_pages/solar_eclipses/TSE_2024_GoogleMapFull.html. Accessed 12 Dec. 2023. (I appreciate this interactive Google map by Xavier Jubier, which allows you to zoom into specific areas to discover the precise start time for each phase of the eclipse).

6 "Great American Eclipse." *Great American Eclipse,* www.greatamericaneclipse.com/. Accessed 12 Dec. 2023.

7 "New NASA Map Details 2023 and 2024 Solar Eclipses in the US - NASA Science." *NASA,* NASA, science.nasa.gov/solar-system/skywatching/eclipses/new-nasa-map-details-2023-and-2024-solar-eclipses-in-the-us/. Accessed 12 Dec. 2023.

8 "National Eclipse." *Eclipse Overview | April 8, 2024 - Total Solar Eclipse,* nationaleclipse.com/overview.html. Accessed 12 Dec. 2023.

9 Jubier, Xavier. *Mexico - USA - 2024 April 8 Total Solar Eclipse - Interactive Google Map - Xavier Jubier,* xjubier.free.fr/en/site_pages/solar_eclipses/TSE_2024_GoogleMapFull.html. Accessed 12 Dec. 2023. (I discovered the San Antonio International Airport resides on the edge of the path of Totality by exploring with this interactive map).

eclipse. They didn't bother to purchase special eclipse glasses, but they did study the shadows on the ground.

Few Experience Totality

During a Total Solar Eclipse, humanity can be divided into two camps: those who experience Totality and those who don't. The path of Totality is narrow. Some are fortunate to live in the path. The time and place of their life will naturally fall under the shadow of a Total Solar Eclipse. In 2017, it was estimated that 12.25 million people lived in the path of Totality.[10] Given that the world's estimated population that year was 7.6 billion, only 0.16% happened to live in the right place at the right time to experience Totality. Sadly, not everyone who lived in the path went outside to look up so they missed it. Others, like my family, traveled to get into the path. It's estimated that two to seven million people, or approximately 0.06% of humanity, went out of their way to journey into Totality for the Total Solar Eclipse of 2017.[11]

Humanity can be divided into two camps: those who experience Totality and those who don't.

The Path of Totality and The Story of the World

A strong parallel between the Total Solar Eclipse and the story of the world is the narrow path of Totality and the narrow path to God. Jesus said, "I am the way and the truth and the life. No one

10 "Statistics." *Great American Eclipse,* www.greatamericaneclipse.com/statistics. Accessed 4 Nov. 2023.

11 Ngeni, Frank, et al. "The 2017 Total Solar Eclipse in the United States: Traffic Management and Lessons Learned." *Transportation Research Interdisciplinary Perspectives,* Elsevier, 9 Dec. 2021, www.sciencedirect.com/science/article/pii/S2590198221002153. Accessed 4 Nov. 2023.

comes to the Father except through me" (John 14:6). In Matthew 7:13-14, Jesus said, "Enter through the narrow gate. For wide is the gate and broad is the road that leads to destruction, and many enter through it. But small is the gate and narrow the road that leads to life, and only a few find it."

Jesus is the only way to God, to eternal life. Some may call this exclusive, but the wildly inclusive truth about the narrow path is that everyone is invited. It doesn't matter who you are, how smart you are, how much money you have, what you look like, where you've been, or what you've done. In the narrow path, none of these distinctions exist—"Christ is all, and is in all" (Colossians 3:11). Interestingly, this beautiful inclusivity makes the path exclusive. What do I mean by that? No drop of hatred is allowed. The path is set apart from evil. It's holy.

Some may call this exclusive, but the wildly inclusive truth about the narrow path is that everyone is invited.

God's Holy Path

God is holy (Leviticus 20:26). The Hebrew word for holy is *qôdesh* (pronounced ko'-desh), meaning to be set apart, separate, sacred, consecrated.[12] God is love (1 John 4:16). He's also patient and kind. He doesn't envy or boast. He's not arrogant. He doesn't dishonor people, not even behind their backs. He's not self-seeking or easily angered. He keeps no record of wrongs. He doesn't delight in evil but rejoices with the truth (1 Corinthians 13:4-6). He's totally good, without even the slightest hint of an ugly, rude, or wrong thought ever.

We humans are not holy like that. Our love is imperfect,

12 "H6944 - qôdesh - Strong's Hebrew Lexicon (NIV)." Blue Letter Bible, Mar. 1996, www.blueletterbible.org/lexicon/h6944/niv/wlc/0-1/. Accessed 13 Dec. 2023.

conditional, and defective. God will not make himself unholy by uniting with us in our unholiness (we discussed this in Chapter 11). Don't we want a God like that? A God who is inviolable, incorruptible, and indestructible? We want a God who is not susceptible to violence or perversion. We long for a God, a leader, a warrior, a hero, who loves without blemish, without even a hint of a hateful, unloving, or perverted thought toward us. We can worship no other and be satisfied. When we taste his holiness, his utterly otherly goodness, we long to unite with him and be holy like him. But before we desired this, he did.

We long for a God, a leader, a warrior, a hero, who loves without blemish.

Did you know that the night Jesus was arrested, the day before his crucifixion, he prayed for every person who would enter the narrow path, that is, for every person who would trust him? He did (John 17:20). We can understand his desire for us by reading his prayer to his Father. People pray for what they long for, and Jesus is no different. He prayed, "Sanctify them by the truth; your word is truth. As you sent me into the world, I have sent them into the world. For them I sanctify myself, that they too may be truly sanctified" (John 17:17-19). To be truly sanctified is to be holy. God wants that for us. He also wants to unite us unto himself.

We see God's desire to be united with us as we continue to read what Jesus prayed the night before his death on the cross. He agreed with his Father's will, "May they [those who believe in me] also be in us so that the world may believe that you have sent me. I have given them the glory that you gave me, that they may be one as we are one—I in them and you in me—so that they may be brought to complete unity" (John 17:21b-23a). Wow, *complete unity!* And did you catch the phrase about the world? "That the world may believe that you have sent me." Jesus was thinking about the whole world, not wanting anyone to perish (2 Peter 3:9). God's heart is for no one to perish, so he made a way for us not to perish. John tells us so in John 3:16: "For God so loved the world that he gave his one and only Son, that whoever believes in him shall not perish but have eternal life." Believing in Jesus is the only way to eternal life. To believe in Jesus is to set yourself apart and be counted holy; it's the only way to unite with God.

Though many religions claim to pave the way, only the person of Jesus can take you to the creator of the Earth, Moon, and Sun, as my brother in Christ, Yusuf,[13] discovered.

Yusuf Enters the Path

Samuel,[14] a missionary in a Middle Eastern country, woke before sunrise to pray, "God, I'm here for your purpose. I have a schedule for this day, but what would you like me to do?"

One unexpected word entered his mind—beach.

Although going to the beach didn't fit Samuel's plan for the day, he obeyed. He adjusted his schedule and made time to go to the beach and walk along the shore. There, he met Yusuf.

Yusuf had been walking the beach for hours, seeking God with a question. Raised a Muslim in the Middle East, Yusuf was acutely familiar with the divide between Sunnis and Shiites. Both groups share the same Islamic beliefs. Muslims believe in one God. They believe Jews and Christians corrupted the stories of God in the Torah and the Bible, but all messages from God have been recorded with divine perfection by Muhammad in the Quran. They also believe in life after death in paradise or hell based on a person's good or bad deeds.[15] So, how do Sunnis and Shiites disagree?

13 Yusuf is not his real name. A pseudonym is chosen to protect this man from persecution.

14 Samuel is not his real name. A pseudonym is chosen to protect this man from persecution.

15 Sherif, Ashraf. "Belief: Six Pillars." *IslamOnline,* 12 Oct. 2023, fiqh.islamonline.net/en/belief-six-pillars/. Accessed 9 January 2024.

The Sunni-Shia divide was birthed at Muhammad's death. Muhammad had no male heir, nor had he chosen a successor for himself. Some followers believed Muhammad's closest relative should take his place and chose his cousin and son-in-law, Ali, as their new leader. This group became known as Shiites.[16] Others believed the leader did not need to be a descendant or relative of Muhammad but should be elected based on his deeds. They chose Muhammad's closest companion, Abu Bakr, as their leader and became known as Sunnis.[17]

Yusuf, longing with all his heart to be right with God, wanted to know which group was in God's will. Samuel met Yusuf on the beach, and the two instantly connected and became friends. Yusuf opened his heart to Samuel as they walked along the edge of the waves. Yusuf said, "I want to be right with God. I'm here to ask him, 'Which way is your way—the Sunni way or the Shia way?'"

Samuel shared another way, the only way, on the sands of the shore that day, and Yusuf joyfully received it as the way of life. He met with Samuel weekly to study the Bible even though he had to do so covertly in a country lacking religious freedom. Eventually, Yusuf's family joined him in Bible study. They entered the narrow path. Today, Yusuf and his family follow Jesus and secretly lead Bible studies in their home. They remain in the narrow path and abide with God under the shadow of religious persecution.

In the path of Christ, we can know God's provisional love under the shadow of every trouble.

As we must remain in the path of Totality and look up to experience the glorious light show of the eclipse, we must also remain in the path of Christ and look up to

16 Newman, Andrew J.. "Shi'i". Encyclopedia Britannica, 9 Jan. 2024, www.britannica.com/topic/Shii. Accessed 9 January 2024.

17 Britannica, The Editors of Encyclopaedia. "Sunni". Encyclopedia Britannica, 8 Jan. 2024, www.britannica.com/topic/Sunni. Accessed 9 January 2024.

God to experience his beauty and goodness. In the path of Christ, we can know God's provisional love under the shadow of every trouble. My friend Elsa is a testament to this truth.

Elsa Remains in the Path

My friend Elsa has witnessed God's provisional love under the shadow of the most heartbreaking and awful circumstances. I asked my sister in Christ to share her story because it's a testament to the truth that no wound is too deep to eclipse God's beauty. The Sun shines through the deepest craters of the Moon to create a diamond ring so amazing that it moves the deepest part of the human soul. God shines his light through Elsa's deepest hurts so beautifully that it gives a strong and confident message of hope to those who listen: Love can be known amid the most tremendous suffering. Joy can follow the darkest darkness.

Elsa's Story
by Elsa Ezell

I remember this day as if it was yesterday. It was 2010, another hot summer morning. My daughter Britta was poolside with her best friend Sam. They were settling in for a relaxing girls' day, dreaming about the next chapter of their lives. Both girls had just graduated and were about to head off to college. Although they were headed in different directions, they spent this day together, laughing and sunbathing.

I sat inside at the breakfast table, praying for my family. I remember looking up and thinking, *What a great life!* Little did I know that our lives would take a drastic turn so quickly.

During this sweet, quiet time, I listened to "Give Me Your Eyes" by Brandon Heath. I turned the lyrics into my prayer, "Lord, show me what I'm not seeing. Give me your love for humanity. Give me your arms for the broken-hearted and those far beyond my reach. Give me your heart for the ones forgotten. Give me your eyes so I can see and love like you do."

When it was time to drive our girl to Northern Arizona University, we had two cars packed. Dad and brother drove one car while Britta and I took the other. Upon arrival to the campus, we learned the school temporarily assigned Britta to a sorority dorm rather than the dorm intended for new incoming students. Things weren't going as they should. Britta assured us she would be fine. But things did not go fine.

Later that night, she attended a sorority party where someone dropped a pill in her drink, and our daughter was raped. She did not want to come home and end her dreams of college, so she kept secrets, not telling us her world was spinning out of control. She started drinking to numb her pain. Then, she was introduced to drugs. By the end of her first semester, we lost our daughter to the world of drug addiction and human trafficking.

I sat in my car in my driveway, screaming at God.

> I KNOW YOU! I KNOW YOU CREATED THE HEAVENS AND THE EARTH. I KNOW YOU CAN SAVE MY DAUGHTER WITH ONE WORD. WITH A SPOKEN WORD, YOU ARE ABLE. I WILL NOT MOVE FROM THIS SPOT UNTIL YOU SAVE MY DAUGHTER!

I sat in complete silence, not moving, determined to see God answer me. I don't know how long I waited. I needed him to answer me.

Finally, I heard a calm, still voice, "If I don't save her, can you still say I AM good?" My heart stirred from within, and my mind filled with all the memories of his faithfulness, the faithfulness of him who has called us into fellowship with his Son, Jesus Christ our Lord (1 Corinthians 1:9). It was his love that kept him on the cross (Luke 23:34-46). Then a peace and assurance that was steadfast in me was resolved at this moment; I would trust this love and lean into him (Proverbs 3:5-6). This moment was just the beginning of our family's six-year journey with my daughter into the dark world of drugs, prostitution, and human trafficking. I felt pain so deep at moments I thought I would not wake up the next morning.

Pain has a way of saturating space.

Pain has a way of saturating space, making one feel like there's no room outside its overwhelming pressure. Those of us who have walked in the depths of its grasp have to admit it's no illusion! The heart cries, "How long must this go on, LORD? Will my flesh be crushed under the pressure of this weight?"

When I was in this space of deep pain, with tears that overwhelmed me, I looked up and set my gaze on the one who could carry and comfort me. Only the Lord, my creator, is enough (Isaiah 43:2). I needed him. Looking up to him in this space is where I found the strength to declare God is good regardless of my circumstances (Psalm 143:8)! I must declare!

It is in declaring that my God is good that I am strengthened. I lean into trusting him, regardless of the outcome. God works all things for good for those who love him (Romans 8:28). I still hurt, have fear, cry, and have to lean into him. But I am getting stronger daily. I am reminded that God is for me, and the enemy does not completely control all the space around me. The saturated space around me is saturated by God (Colossians 1:16)!

For the next six years, I walked with my daughter in this world of deep pain. I was reminded of the day I prayed for the Lord to give me his heart, his arms, his eyes, his love. As I got to know the other homeless prostitutes and drug addicts, I saw they were also sons and daughters, likely loved by another mother or father and definitely loved by God. How did they get here? How is it I never cared?

Evil is always lurking to steal, kill, and destroy (John 10:10, 1 Peter 5:8). Human traffickers take steps to trap their victims. First, they befriend, then they intoxicate, alienate, isolate, desensitize, and finally capitalize.[18] But the light shines in the darkness, and the darkness has not overcome it (John 1:5). We walk in his light and do not need to walk in fear (2 Corinthians 4:6).

The love of God is consistent. It's always been around us and continually pursues us.

The love of God is consistent. It's always been around us and continually pursues us. When we take time to study the galaxies, we begin to see how grand and complex this world is and how little we know. Love is like this. We think we know what love is because we've experienced human love. But God's love for us is like the galaxies. Once you begin to grasp the reality of who God is and his love for you, it changes how you see things and how you live your life. God's love is chasing after you. He doesn't love us because we deserve it or earn it—that's humanity's way of loving (Romans 5:7-9). While we reject and offend God, Jesus said, "Forgive them, for they know not what they are doing" (Luke 23:34). God delights in showing mercy (Micah 7:18).

18 Unanue, Bob, and Ally Brito. *Goya Cares Educational Series: "LIGHT." YouTube,* YouTube, 2 Mar. 2022, https://www.youtube.com/watch?v=Rx3tmPkDijo. Accessed 5 Nov. 2023.

On Galaxies and God's Love

Elsa said, "God's love is like the galaxies," which reminds me of the remarkable discoveries of the Hubble Space Telescope. Dr. John Mather, Senior Astrophysicist at NASA, said, "In astronomy, there's what we knew before Hubble, and now there's what we know after Hubble." He goes on to explain that Hubble's discoveries significantly changed astronomy textbooks.[19] Hubble marks "the most significant advancement in astronomy since Galileo's telescope," and its discoveries are the "backbone for more than 15,000 technical papers."[20] Why? The advent of Hubble showed us that where we thought there was nothing, there are indeed many somethings.

Hubble, a telescope that circles our blue marble 15 times a day, has a major advantage over telescopes on Earth.[21] It doesn't have to peer through the veil of Earth's atmosphere to get a clear picture of space. Five years after its launch in 1990, scientists pointed Hubble at a small dark patch in the sky near the handle of the Big Dipper, also known as Ursa Major. In December 1995, Hubble took photos of this seemingly empty spot for ten days, including Christmas Day. The resulting image was like a surprising and wonderful Christmas gift. In this tiny pocket of what looked like nothing, almost 3,000 galaxies were discovered.[22] Given that we estimate each galaxy contains 100 billion stars, that's 300,000,000,000,000 or 300 trillion stars in a spot that looked like nothing![23]

---

19 Morris, Paul. *The Hubble Deep Field: Looking Back In Time. YouTube,* NASA Goddard, 2 Aug. 2021, www.youtube.com/watch?v=Gr_AF_AB1AU. Accessed 5 Nov. 2023.

20 "Hubble Space Telescope - NASA Science." NASA, NASA, science.nasa.gov/mission/hubble/. Accessed 5 Nov. 2023.

21 "Hubble Observatory - NASA Science." NASA, NASA, science.nasa.gov/mission/hubble/observatory/. Accessed 5 Nov. 2023.

22 *"The Hubble Deep Fields." ESA/Hubble,* ESA/Hubble, esahubble.org/science/deep_fields/. Accessed 5 Nov. 2023.

23 Jackson, Brian. "Astro for Kids: How Many Stars Are There in Space?" *Astronomy Magazine,* 18 May 2023, www.astronomy.com/science/astro-for-kids-how-many-stars-are-there-in-space/. Accessed 5 Nov. 2023.

In 2004, scientists pointed Hubble to another empty spot, this time in the constellation Fornax, also known as the Furnace.[24] After studying this area for years, Hubble discovered approximately 10,000 galaxies in this tiny speck.[25] Wow! That's a lot of stars in what looked like nothing.

I agree with Elsa. God's love is like the galaxies. God's love is uncountable, without end. When we think we've experienced God's love to the fullest, he surprises us with more. Additionally, when we're able to see beyond the veil of the troubles of the Earth, beyond the worst of evils such as abuse, alienation, addiction, extortion, and human trafficking, we find God's love pre-existing us, reaching for us, lighting up the dark of seemingly empty spaces in ways we never could have imagined, and our lives are forever changed.

We find God's love pre-existing us, reaching for us, lighting up the dark of seemingly empty spaces in ways we never could have imagined, and our lives are forever changed.

God in the Shadow of Suffering

Elsa sat across from me in a coffee shop in June of 2021, a little over four years after her daughter breathed her last at the young age of 24. With an adventurous twinkle in her eyes, Elsa shared her favorite Bible stories with me and personal stories that make

24 "The Hubble Deep Fields." *ESA/Hubble,* ESA/Hubble, esahubble.org/science/deep_fields/. Accessed 5 Nov. 2023.

25 "Hubble Ultra Deep Field 2014 - Hubblesite.Org." *Hubblesite,* Space Telescope Science Institue, 3 June 2014, hubblesite.org/contents/media/images/2014/27/3380-Image.html. Accessed 5 Nov. 2023.

you smile. She carried a deep joy, which is why I wanted to meet with her. How can someone have such joy in the aftermath of such grief? When I asked her about that joy, she said, "With Christ, there is unimaginable joy, and there is also unimaginable suffering. We see it on the cross."

Later that month, she invited me to a Parents of Addicted Loved Ones (PALS) meeting, a support group for parents with adult children dealing with substance use disorder. That night, she and her husband Charles planned to share parts of Britta's story, and I was welcome to attend and use anything she shared for this book. When I arrived, I sat in one of the ten chairs set up in a circle. I felt overwhelmed knowing each person's presence pointed to the heartache and slavery of addiction. As we went around the circle for introductions, my heart broke for every mother and father, and I silently cried out for each person's child, "Oh God! Please help!"

Elsa and Charles led the group and shared their story. Britta went to rehab and relapsed. Elsa rented Britta an apartment in California and moved in with her to help her achieve her dream of becoming a cosmetologist. Creating a safe routine, Elsa repeated the same activities each day. Elsa made Britta breakfast, drove her to Paul Mitchell The School Costa Mesa, and waited in the parking lot until lunchtime when they ate together. After school, Elsa picked her up and fed her a good dinner. A good mother is willing to lay down her own life for her child's sake, be consistent like the rising and setting Sun in the sky, to watch over, protect, nourish, and provide. Each day followed the planned pattern until one day, it didn't.

Britta didn't come out for lunch. Elsa went in to look for her only to discover she had left out the back door that morning with James,[26] a bodybuilder Britta met in rehab with bulging arms so massive they couldn't lay flat against his sides. James deceived Britta

26 James is not his real name. A pseudonym is chosen to protect his privacy.

into believing he loved her and that they were meant for each other.

Elsa didn't know where James lived, but she knew the snow ski shop where he worked, so she drove there immediately and anchored herself to a chair in the lobby, demanding the boss give her James's home address.

"I don't want this in my business, ma'am. I don't want this trouble with you," the boss asserted.

"Then give me his address, and I'll be gone," Elsa demanded. The man could not compete with the perseverance of a good mother's love and gave her the address, which she passed on to Charles.

Charles immediately drove to the address where James lived amid hallowed-out tanker storage containers repurposed into makeshift apartments. Charles knocked on the door, and James answered. "I'm here to get my daughter," Charles stated like light bursting forth in this dark place.

"She's mine," James lied, "She no longer belongs to you. She belongs to me."

Charles disagreed, and a physical fight ensued. After taking an onslaught of punches, Charles rose from the ground. With the little breath he had left, Charles spoke again, "Give me my daughter."

James stepped aside and led Charles upstairs. Charles searched the house until he found his girl drugged in the shower. She was so out of it she didn't want to go and only weakly repeated, "Hey, Dad," in a voice not her own. Turning off the water and wrapping her in a towel, he gently led her, "Come on, let's go," and walked her out the door.

After months of recovery work, Britta relapsed again. The story hurts. I asked God, "Where were you?" I heard God answer: *I was in her father, Charles, who went after his daughter and, after being beaten up, rose again with words no man could stop, "I'll go get my daughter now."*

A woman at the PALS meeting, sitting on my left, leaned forward with eyes set on Charles and Elsa. "Tell me about the joy on this side of it. I knew joy before. Tell me about the joy after." She recognized what I recognized—joy, deep and strong in this steadfast couple under the worst of shadows. It's awe-inspiring.

The situation did not end as Charles hoped. Charles dreamed Britta would stand on stage with a microphone, "Boy, have I got a story to tell," she might begin, "God transformed my life from tragedy to triumph." But many friends Britta met in rehab stood on stage and spoke these words at Britta's funeral.

An ex-prostitute called Elsa when she heard of Britta's passing. "Britta saved my life," she said. She shared a memory of sitting on a bench. Britta came and sat alongside her and told her about Jesus, her beauty, and how she didn't belong there on that bench, prostituting herself. Britta's words also shine in written form.

Britta's mother-in-law discovered Britta's journal in June of 2021, the month I attended the PALS meeting. She mailed it to Elsa. A few weeks later, Elsa excitedly called me, "Amy! Britta wrote about an eclipse!" She texted me a picture of the relevant page she found in Britta's journal. Britta wrote, "sun-moon (alignment) we must stay in alignment with God."

So...What's the Point?

The point is that there is only one narrow path from which you can experience the Total Solar Eclipse, and there is only one narrow path to God. Jesus is the narrow path, and you're invited to enter. While you may need to travel 1,000 miles to get in the path of an eclipse, as I did in 2017, you don't have to travel at all to enter the path of God.

You can enter right where you are: at home, at work, in a store, in a car, in a chair, on a walk, in a bathroom, anywhere. We enter the narrow path by trusting Jesus. We can remain there to experience his goodness and glory under the shadows of many troubles.

I hear God's voice in the Total Solar Eclipse, the glorious light show that speaks to the deepest part of our beings. We feel its message in our bones. Awestruck, we can't help but declare, "Oh my God!"

We can put the message of the eclipse into words. Can you hear it with me?

I hear God whisper in First Contact: *I see your need for salvation. I recognize the darkness that attempts to take over the Earth and your heart. Don't be afraid, you're special to me. I made you with pleasure on purpose because I want you. I love you.*

I hear God whisper in the diamond ring of Second Contact: *It's OK if you fail. I will keep my promise and pursue you. Nothing can separate you from my love.*

I hear God whisper in the pearly crown of Totality: *I wrap myself in flesh that you might know me, my extravagant love, grace, justice, and power to conquer death. No darkness, no evil can eclipse my love for you. I will die for you. I have died for you. And what's more, I have come back to life for you so you can live with me forever.*

I hear God whisper in the diamond ring of Third Contact: *Only put your trust in me. I am a promise keeper. I will be with you forever. I desire to bring you freedom, to separate you from every wrong thing, and to make everything right. I will do it. I am doing it.*

I hear God whisper in Fourth Contact: *My light is spreading. Rejoice! The darkness cannot overcome it. I will stay with you in every trouble. Stay with me.*

God's extravagant love exceeds our expectations when we discover it. God exceeded my expectations throughout the journey of creating this book, all the way to the very end.

God's extravagant love exceeds our expectations when we discover it.

Days before this book was due to the publisher, a friend told me, "I'm not sure the title is right. The word *glory* is vague, and I'm not entirely sure what it means." I agreed. My daughter is named Glory, and we often ponder the broad meaning of her name. I chose her name because I knew glory meant something good. The word appears 300 times in the Bible with multi-faceted meaning.[27]

We feel every meaning of glory when we stand in the shadow of Totality and witness its fiery halo.

We feel every meaning of *glory* when we stand in the shadow of Totality and witness its fiery halo. It comes with thoughts of honor, might, power, kingship, and victory in battle. Psalm 24:8 comes to my mind, "Who is this King of glory? The Lord strong and mighty, the Lord mighty in battle."

Glory also comes with the notion of beauty and magnificence, "All people are like grass, and all their glory is like the flowers of the field; the grass withers and the flowers fall, but the word of the Lord endures forever" (1 Peter 1:24-25a).

Furthermore, *glory* lends itself to a sense of otherness, deep awe, reverence, shining light, and terror, "An angel of the Lord appeared to them, and the glory of the Lord shone around them, and they were terrified" (Luke 2:9).

Moreover, *glory* goes along with ideas of praise and adoration. "Glory to God in the highest heaven, and on earth peace to those on whom his favor rests" (Luke 2:14).

Additionally, *glory* conjures feelings of reverence and weight, a mysterious greatness we cannot fully understand or face. I think of Moses when he asked God, "Show me your glory," but God said, "You cannot see my face, for no one may see me and live" (Exodus

27 "'glory.'" *Blue Letter Bible*, www.blueletterbible.org/search/search.cfm?Criteria=glory&t=NIV#s=s_primary_0_1. Accessed 20 Jan. 2024.

33:18-20). Yet God wrapped his glory in flesh, that we might look right at him.

John 1:14 says, "The word became flesh and made his dwelling among us. We have seen his glory, the glory of the one and only Son, who came from the Father, full of grace and truth." We have seen his glory.

And we have seen his story and ours in the Total Solar Eclipse, his pursuit of us from fall to restoration. I thought of changing the name of this book to *The Story in the Total Solar Eclipse,* but my husband disagreed. Sitting at the kitchen table, our family began to contemplate the title.

My gut pulled me toward the mystery, weight, and manifold meaning of the word *glory.* After all, this book is wildly varied, like the word's meaning. We've covered space science, faith, philosophy, and honest, delightful, terrifying, and awe-inspiring stories. Still, my mind felt drawn to the simplicity of the word *story* until my daughter Glory sealed it for me.

She typed something in her phone and then looked up, smiling with a twinkle of confidence in her eyes as she announced, "The fourth definition of glory is: 'a luminous ring or halo, especially around the head of Jesus Christ or a saint.'"[28]

Once again, the eclipse, in all its glory, points us to Jesus, the one who shares his beauty, bestowing on his people "a crown of beauty instead of ashes" (Isaiah 61:3).

Jesus, in all his generosity, compassion, holiness, goodness, love, kingship, and power, *is* the glory of the Total Solar Eclipse.

28 "Glory, N., Sense 9.a." Oxford English Dictionary, Oxford UP, September 2023, https://doi.org/10.1093/OED/7726517090. Accessed 20 January 2024. (Glory's Google search brought up a dictionary box served by Oxford Languages. I purchased a one month subscription to the official Oxford English Dictionary to confirm this definition. I found two definitions that serve the point well: Definition 9.a "the circle of light represented as surrounding the head of the whole figure, of the Saviour, the Virgin, or one of the Saints," and Definition 9.c "Any circle or ring of light; a halo, corona.")

WITH GRATITUDE

So many people poured into this book that my acknowledgments "page" has become the size of a chapter. Although a book is written through countless hours in a solitary place, it is certainly not written in solitude.

My amazing daughter and illustrator, Joy Burgin Fish, thank you for illustrating this book. A book with pictures is the best kind, and I especially love these illustrations because they give us a glimpse of your astounding inner beauty and talent. You make my dreams come true. Thank you for being so wise and good at serving people in such a way as to bring out their best. You do that for me and many others. Thank you for bringing to fruition what I longed for in a book cover but couldn't verbalize—you had me squealing with delight for hours.

To my editor, Jill E. McCormick, thank you for taking me on. Your brilliance and beauty bring me to tears when I think of working with you. You were not afraid of or appalled by my mess. Each time you gave yourself to this work, you were like the Spirit of God hovering over the chaos of beginnings. You saw the good and called it forth. Knowing my heart, the heart of our God, and the needs of our readers, you served us all deeply and well. I am forever grateful.

Mandy Pallock, thank you for taking me on. You are an answer to prayer, and I am amazed by God's generous provision, meticulous care, and beauty given to me through you. You treasured me and my ideas when we met at the coffee shop to initiate the plans to get this book printed and published. You have designed a gorgeous book.

Elsa Ezell, thank you for encouraging me and believing this book would one day exist when only bits and pieces were spread out in various places like leaves scattered on the ground. You sent me notes and stickers encouraging me that the dream was meant to come true. To you and Charles, thank you for allowing me to include your beautiful and steadfast hearts in the last chapter of this book. May God's kingdom of light spread and his people soar because of it.

Bob Unanue, thank you for your kind endorsement and heart of leadership in our world. May the light of God continue to spread through Goya Cares.

Danielle and Mario Lopez, thank you for allowing a stranger to step into your lives and include glimpses of your beauty in this book. Your transparency and grace are strong and beautiful teachers to me and countless others. Danielle, thank you for your continued encouragement and texts as the book neared its completion and for quickly becoming a dear friend.

To my first writing coach and dream defender, Renee Fisher, thank you for hosting the first writer's workshop I ever attended. You passionately encouraged me when I sheepishly showed up with my first bits and pieces of this book. You took a picture of me holding the page with the draft title. The photo was my first glimpse that this idea could become a part of the physical world.

To my second writing coach, Ann Kroeker, I came to you with a big mess of an idea I couldn't seem to let go of or wrangle. It felt like a wild bull, and you helped me lasso it. Your *The Art & Craft of Writing* course helped me connect more deeply with my love of writing and provided the space where this book began to take its final shape. You strengthened me to push back against the resistance that comes against every writer by pressing me to "do the work." Further, it is you (and Glory) we can thank for the concluding sections of each chapter titled *So... What's the Point?* We needed those. Every interaction with you came with many gifts that contributed to this book's development. Thank you.

Jenny Fish and Nicole Standiford, thank you for taking the time to read this book and provide valuable feedback. The book is better and more kind because of you.

To the authors I've learned from yet never met such as Emily P. Freeman, Gary Morland, Daniel Nayeri, Dani Shapiro, Beth Moore, Anne Lamott, and more. Thank you for mentoring me from afar with your writing.

Laurie Ogier, I will love you forever for showing me Jesus and inviting me to church. It seems every time I give a glimpse of my testimony, you are introduced.

I thank CrossBridge Community Church for teaching and growing me for almost two decades. Kirk and Debbie Freeman, thank you for starting and leading our church with Jesus at center stage. Your faithfulness has given God much room to change us and our legacy.

To the Todd and Krystle Newton Life Group, thank you for praying for my family and me as I wrote this book in the face of the tragedy of my father's stroke and all the other trials only you, my immediate family, and closest friends know. Maybe I can find time to cook for us again soon.

Sara Branstetter, JoLynn Posey, and Brenda Hanson, you were the first friends I confided in that I had a desire to write a book. Strangely, I felt ashamed to tell anyone, but Sara, you looked me straight in the eyes and spoke as if you had the authority of the King of the Universe, "Amy, there is no shame in writing a book." JoLynn and Brenda, you agreed without hesitation. That truth settled deeply into my soul and compelled me forward. I want to be like that, to speak as rightly to people as you did to me that day.

Stephanie Aaron, Stacy Neighbors, Marian Hall, and Jeff Gonzalez. Each of you spoke a personal word to me that directly contributed to the strength required to finish this book. None of you knew of my project or the way your words helped me finish.

Stephanie, you could not have known how I was wrestling with the size of this project, still blind to the structure of this book when you said, "Don't be afraid of the size of the task. God is with you."

Stacy, you said, "You can bloom in the desert." Knowing I can bloom under the hot, dry winds of any trial has spurred me on in many areas of life. Our environment doesn't matter because our roots go deep.

Marian, you spoke over me when I wondered if quitting would be wise, "Be faithful to the end," you intently repeated without specifically knowing how the words applied. These words brought me to my knees in prayer until I heard God speak them to me as a creation word. God said "Be faithful to the end," so I got up and was. Thank you.

Jeff, you said, "God is pleased with your tenacity." Wanting nothing more than to bring a smile to his face, your words encouraged me to keep my grip no matter what.

Brad Trantham, thank you for hiring me and for allowing me to include my hiring story in this book. I thoroughly enjoy working with you and the entire team. I have been given a close-up view of the strong way you care for each person on our team. It's beautiful and admirable.

Rob Thorpe, thank you also for hiring me and allowing me to include my hiring story in this book. You have built a special team in support of space science, and I am thrilled to be a part of it and help it thrive. Thank you for taking the time to carefully read this book in advance and provide valuable feedback. Our readers can thank you for smaller paragraphs and a better book. Most importantly, you have encouraged and prayed for me like a brother.

To Carole Anne Payne, thank you for being such a dear friend to me all these years. Thank you for all the walks around our neighborhood, for praying for this book, for brainstorming ideas with me, and for supporting me in my endeavors. You cheer me on despite knowing so many of my flaws.

To Rebekah Payne, I must acknowledge someone as

thoughtful, cheerful, and encouraging as you. Thanks for asking me how the book was going and for stopping by when walking past my house. Those little surprise visits are the best. And thank you for telling me you recently read a book whose acknowledgments were so grand, they formed a chapter of their own. I'm afraid I have done the same.

To Jon's parents, you gave me the gift of the eclipse, the little seed that started the journey of this book. Most importantly, you have given the world the gift of your son, Jon. I am forever grateful. Thank you.

To CJ Burgin, thank you for your gift of hospitality and for coordinating our trip to the 2017 eclipse. You provided the eclipse glasses and told us when to look up. I will never be the same because of it.

To Debbie Burgin, thank you for sending me a copy of the beautiful photograph of the corona you captured in 2017. I kept it on my bulletin board above my computer as a reminder of its beckoning to explore its beauty. Also, you were a kind reader of my blog when I timidly began the journey of writing publicly. At times, you sent me thoughtful emails in response. I loved that.

To Nanny, I miss the days when we could stay up and talk until morning. Thank you for loving me and pointing me to Jesus with your life.

To my phenomenal parents, thank you for giving me the strong foundation necessary to complete any work. You are overcomers, and I am forever proud of you. I hear your voices in mine and work to carry forward all the good ideas I learned from you. Being secure in your love has made it easy for me to find security in God's love.

To my lovely cousin, Candi, thank you for setting a timer to pray for me every Monday, Wednesday, and Friday at 12:43 p.m. Thank you for staying connected to me through our plethora of Marco Polos. You have been exposed to my many wild thoughts and have loved me anyway. You have loved me with words of affirmation, gifts, and trips to San Antonio. Together, we are Colossians 3:16 in the flesh.

To Philemon, the surprise arrow in my quiver, thank you for calling me your mother. I am proud to call you my son. You are a bright light in Rwanda. Where the country suffered under the darkness of hate speech broadcast on the radio in the 1994 genocide (the year you were born) the country now benefits under the light of your voice broadcast on the same radio waves. The alcoholic puts down his drink. The suicidal stops her attempt. The hopeless find hope, lift their heads, and persevere. I am incredibly proud of you.

To Joy, Glory, and Freedom. You have been the cloud of beauty surrounding me everywhere I went over the last two decades. Now you are spreading out on Earth like a stunning landscape before me. I am amazed by the view. Nothing I have seen surpasses your beauty. Somehow, this universe in all its vastness is not as grand as you are.

Freedom, I loved having my office space right across from your bedroom this year. When I wrote in the middle of the night, we often bumped into each other, and you cheered me on. With your hands steady on my shoulders, you looked me straight in the eyes many times, "You can do it, Mama." You point me to Jesus when I need him, and for that, I am forever grateful. You are calm and steady when you bail me out of stressful situations (like when I drive over traffic cones, lodge them in the undercarriage of the car, and freak out). As you lean into your name, I think the key resides in 2 Corinthians 3:17b, "Where the Spirit of the Lord is, there is freedom." I have seen God's hand upon you and your response to him—he makes you smile, which makes me smile.

Glory, your perseverance to finish many creations spurred me on when I was tired. I thought of you often knowing you would never quit but push through to the end. You were an example to me. How could I give up with a daughter like you watching? I have loved having you home this last year of college. Thank you for putting your brain on ideas with me. I sit in my office surrounded by the many items you have created for me, mostly lately, the little mouse Stewart that you painted above the baseboard, and I replay your voice encouraging me, "Mom, this section has ethos, pathos, and logos in it—good job!" I replay your voice helping me push through my weakness, "Mom, you have to make a point. So, what's the point?" I hope I have made one. Thank you. Now you have brought me another son—Austin, thank you for your hugs, your smile, your thoughtfulness, your right-thinking, for washing the dishes, and loving my daughter.

Joy, God has given me such artistic daughters! The 3-D nightlight you made with the deer standing under the eclipse and stars stood as a mandate before me to finish this book. The little lightbulb with flowers inside reminded me to do with an idea what I hope my children would do with their ideas. How could I not finish when my girls gift me with such creations? Thank you for partnering with me in the Christgazing podcast and the creation of this book. And, you have brought me another son—Ethan, you are a reliable rock, hard-working, thoughtful, and true. Thank you for loving Joy and embracing us, though we are about as wild as The Croods.

Jon, you joyfully created and protected space for me to write. You read every word and corrected me to ensure sound logic and flow. When I am confused, you are not. When I am afraid, you are not. You fill my gaps with a goodness that exists nowhere else but in you. You are the one I rely on and lean on like no other. You are the one who knows my faults like no other and yet serves and loves me like no other. Thank you. If I am able to accomplish anything, it is because God has given me you, my sure and steady earthen rock, my provider, protector, and more. Without you, I am so much less than I am.

Lastly, I would like to thank my Lord, the creator of everything beautiful and good. I love you. May my entire life be a living book of gratitude and acknowledgment to you. Just say the word over me, and I will be.

ABOUT THE AUTHOR

Amy Burgin is a computer scientist at Southwest Research Institute, where she leads and serves multiple teams in support of NASA space missions. She has been actively involved at CrossBridge Community Church since 2006, where she has volunteered in many leadership roles, such as coordinator of the pre-K program, author of a weekly devotion for children, director of 5K Race for Rwanda, and speaker. Living in Texas with her supportive husband, she enjoys writing, laughing, learning, and spending time with her beautiful adult children. She also co-hosts the Christgazing podcast with her daughter, Joy.

Connect

amyburgin.com
instagram.com/amyburginwriter

ABOUT THE ILLUSTRATOR

A journalist by day, Joy Burgin Fish considers herself a jack of all trades. She is an anchor and producer for two local TV stations, News 4 San Antonio and FOX SA. She also co-hosts the Christgazing podcast with her mother, Amy. A few of her other hobbies and professions include painting, singing, photography, voice-over work, and competitive figure skating. When she's not doing any of that, she enjoys spending time outdoors with her hard-working and supportive husband and their one-eyed cat.

Connect

joyburgin.weebly.com
instagram.com/joy.burgin.tv

When you buy this book, all sales go directly to Meant to Soar, a 501(c)(3) nonprofit organization founded by Elsa and Charles Ezell, parents of Britta, whose story is shared in Chapter 13.

Meant To Soar is dedicated to breaking the cycle of substance abuse and human trafficking by identifying children with wounded hearts and equipping them to be a strong generation and know their true identity.

Visit meanttosoar.org to find out more.

the Christgazing podcast

Do you want to know Christ more, or learn who Jesus is for yourself? Join Amy Burgin and Joy Burgin Fish each week as we create space to focus on his word and listen for his voice. If you desire God's presence and holy direction in the midst of your daily life, this podcast is for you.

Listen to the Christgazing Podcast wherever you get your podcasts!

www.ingramcontent.com/pod-product-compliance
Lightning Source LLC
Chambersburg PA
CBHW070942100426
42737CB00011BA/1564

* 9 7 9 8 9 8 8 4 7 8 9 3 5 *